CONTENTS

Department of Trade and Industry

Energy Paper Number 62

LONDON: HMSO

ISBN 0 11 515384 5

ENERGY PAPERS

This publication is the 62nd in the series of Energy Papers previously published by the Department of Energy. This paper is the third in the series to be published by the Department of Trade and Industry.

The series is primarily intended to create a wider public understanding and discussion of energy matters, though some techinical papers appear in it from time to time.

The papers do not necessarily represent Government or Departmental policy.

Other papers in the series which began in 1975 can be indentified on request from the Librarian, Department of Trade and industry, 1 Palace Street, London SW1E 5HE. Telephone: 071-238 3042

The paper used in this publication is produced by an elemental chlorine free process.

PREFACE

1. The purpose of this document is to state the Government's policy towards new and renewable energy and to summarise its strategy and programme for implementing that policy. It results from a review which has considered both independently produced documents such as:

- Energy Select Committee Fourth Report on Renewable Energy (Ref 1);

- National Audit Office report on the Renewable Energy Programme (Ref 17);

and the following specifically commissioned reports:

- Renewable Energy Advisory Group (REAG) Report (Ref 2)

- An Assessment of Renewable Energy for the UK (Ref 3).

The review was conducted against a background of general Government and European Community (EC) policy towards the need for sustainable development and the Government's policy towards science and technology (Refs 5, 6, 7, 8). It has followed on from the publication of the UK programme on Climate Change (Ref 14) and the UK Strategy for Sustainable Development (Ref 15). Additionally, the Assessment of Renewable Energy for the UK (Ref 3) forms the renewables contribution towards the Appraisal of UK Energy Research, Development, Demonstration and Dissemination (Ref 10).

2. This document does not attempt to reproduce information and argument from those reports. Rather, it presents a statement of the Government's view towards new and renewable energy founded upon them.

3. In 1988, Government presented updated plans for the development and exploitation of renewable sources of energy. The strategy and twelve year forward programme were published as Energy Paper 55 (Ref 9). That strategy would have taken technologies with prospects of being economically viable and environmentally acceptable to the point of demonstration assuming that a market would then have taken over. The strategy has been broadly followed, with the addition of the Non-Fossil Fuel Obligation (NFFO) for electricity producing renewables. Orders have been made to help stimulate a market in England and Wales since 1990, and in 1993 proposals for Orders were announced for Scotland and Northern Ireland. Throughout this document the Obligations are collectively referred to as NFFO.

4. There have been a number of other major developments relevant to the energy market in recent years which have caused the strategy and programme to be reviewed. These factors include:

- increasing awareness of environmental issues including the need to take precautionary measures to limit the emission of greenhouse gases, and the prospect for cleaner heat and power generation by using new and renewable energy technologies;

- the changing of the market by the privatisation of the gas and electricity industries, and the planned privatisation of the coal industry, in the UK;

- the move towards a unified European market;

- evolving policy on waste, recycling and use of agricultural land;

- developments in understanding of the energy technologies resulting from research, development, demonstration and dissemination (RDD&D);

- greater international involvement by the DTI's new and renewable energy technology programme.

5. The 1990 Environment White Paper (Ref 7) announced a review of renewable energy, and also indicated that Government would work towards a figure of 1,000 megawatts (MW) of new renewable electricity generating capacity in 2000. In 1993 the Coal Review (Ref 5) concluded that the Government would work towards an increased figure of 1,500 MW for the UK as a whole by the year 2000 and would publish this strategy review document.

NEW AND RENEWABLE ENERGY
Future Prospects in the UK

Energy Paper Number 62

1. THE GOVERNMENT'S POLICY FOR NEW AND RENEWABLE ENERGY

Government policy is to stimulate the development of new and renewable energy sources wherever they have prospects of being economically attractive and environmentally acceptable in order to contribute to:

- diverse, secure and sustainable energy supplies;
- reduction in the emission of pollutants;
- encouragement of internationally competitive industries.

In doing this it will take account of those factors which influence business competitiveness and work towards 1,500 MW DNC * of new electricity generating capacity from renewable sources for the UK by 2000.

2. NEW AND RENEWABLE ENERGY - STATUS

2.1 Current Situation

2.1.1 Renewable energy is energy which occurs naturally and repeatedly in the environment and which can be harnessed for human benefit. The ultimate sources of renewable energy are the Sun, the Earth's rotation and internal temperature, and gravity. The main carriers are the wind and the ocean; wood and crops; the fall of water from lakes and rivers; and animal and human waste including domestic, commercial and industrial wastes. Other new technologies such as fuel cells offer prospects for improving the effectiveness with which energy sources can be exploited.

2.1.2 The extent to which new and renewable energy technologies are already deployed varies widely throughout the world depending upon geographic and climatic features, the comparative economics of other energy systems and the level of development of the individual technologies. Generally they are at much earlier stages of development than other energy technologies.

2.1.3 World wide the market penetration of new and renewable energy is significant, providing about 20% of the world's electricity (Ref 11). This supply is dominated by large-scale hydro power with approximately 550,000 MW of capacity installed world wide. Additionally there are significant and growing contributions from other renewable sources with 6,000 MW electrical capacity from geothermal aquifers and over 3,000 MW of wind capacity. In the United States alone, there is 9,000 MW of biomass fuelled electricity generating capacity, and the use of wood for heating is widespread in the developing nations. Renewable resources including traditional biomass and large scale hydro, currently contribute about 18% to mankind's world energy needs (Ref 12).

---

*Megawatts, Declared Net Capacity (refer Annex 2)

2.1.4 Renewable energy sources provide approximately 6% of the European Community's primary energy production (about 4% of consumption). Biomass, hydro power and geothermal energy provide about 60%, 30% and 6% of that total respectively. Although the use of renewable sources for electricity generation in Europe is dominated by hydro power which provides approximately 8% of the European Community's total requirement, in recent years there have been small but increasing contributions from biomass and wind.

2.1.5 Within the UK, renewables currently provide less than 1% of primary energy and about 2% of the 1992 electricity supply (6 Terawatt-hours per year, TWh/y), the main component being 1,200 MW of hydro-based capacity in Scotland, predominantly from large-scale schemes. Figure 1 shows the breakdown of renewable energy utilisation for 1992. The 1989 Electricity Act paved the way for the Non-Fossil Fuel Obligation (NFFO), and this has provided a powerful impetus to further deployment of the electricity producing renewable energy technologies. By the end of 1993 nearly 200 NFFO projects had been contracted in England and Wales totalling around 620 MW Declared Net Capacity (DNC) though it was not clear how many would finally come to fruition. 134 projects totalling 251 MW DNC had commenced generating. Annex 1 provides further detail of these figures.

2.1.6 Contributions to the overall UK energy supply from solar radiation are substantial, providing lighting and heating to buildings. Unplanned use of solar radiation makes a contribution to energy demand in the building stock, recently estimated to be about 150 TWh/y (Ref 16). There is also a small and underdeveloped heat market supplied by biomass, predominantly wood.

2.1.7 The development of new and renewable energy technologies varies widely in the UK. Most are at an early stage with only conventional large scale hydro operating commercially in a fully established market. Some technologies are competitive in niche markets, however overall the technologies are immature and need developing together with the market and supplying industries if they are to become competitive. A summary of their status for application in the UK market is given in Figure 2 with fuller descriptions in Annex 2 and comprehensive discussions available in the Assessment (Ref 3).

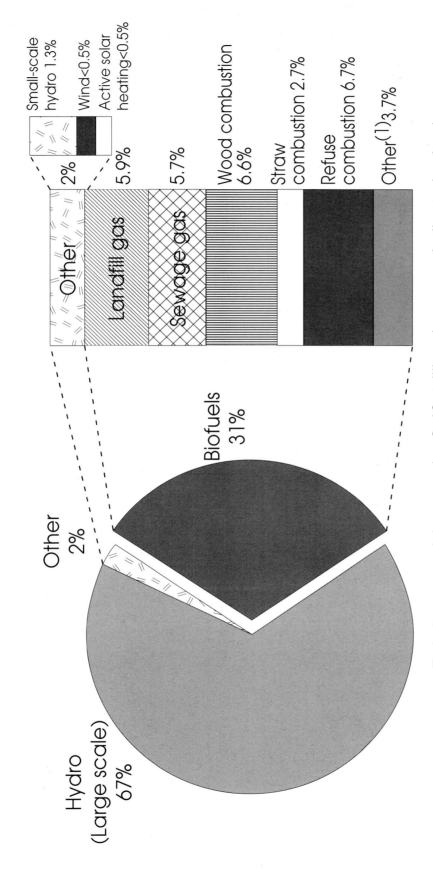

Renewable Energy Utilisation 1992

Figure 1

Small-scale hydro 1.3%
Wind<0.5%
Active solar heating<0.5%

Other 2%
Landfill gas 5.9%
Sewage gas 5.7%
Wood combustion 6.6%
Straw combustion 2.7%
Refuse combustion 6.7%
Other(1) 3.7%

Other
Landfill gas
Sewage gas

Other
2%

Biofuels
31%

Hydro
(Large scale)
67%

Total renewables used = 2.48 million tonnes of oil equivalent

(1) "Other" includes farm waste digestion and chicken litter, industrial and hospital waste combustion.
N.B. This figure excludes unplanned use of passive solar energy, estimated to be about 11.8 million tonnes of oil equivalent annually.
Reference: Digest of UK energy statistics, 1993, HMSO.

Figure 2: Current Status in the UK of the New and Renewable Energy Technologies

	Inappropriate for the UK	Research	Development	Demonstration	Commercially Available	Established Market
Hydro Large Scale					xxxxxxxxxxxxxx	xxxxxxxxxxxxxxxxxxxxx
Simple Passive Solar				xxxxxxxxxxxxxx	xxxxxxxxxxxxx	
Hydro Small Scale					xxxxxxxxx	
Landfill Gas				xxxxxxxxxxxxxx	xxxxxxxxxxxx	
Active Solar					xxxxxxxxx	
Onshore Wind Energy			xxxxxxxxxxxx	xxxxxxxxxxxxx	xxxxxxx	
Specialised Industrial Wastes			xxxxxxxxxxxx	xxxxxxxxxxxxx	xxxxx	
Municipal & General Industrial Wastes			xxxxxxxxxxxx	xxxxxxxxxxxxx	xxxxxxx	
Advanced Passive Solar			xxxxxxxxxxxx	xxxxxxxxxxxx		
Photovoltaics		xxxxxxxx	xxxxxxxxxxxx	xxxxxxxxxxx		
Geothermal Aquifers		xxxxxxxx	xxxxxxxxxxxx	xxxxxxxxxxxx		
Agricultural and Forestry Wastes			xxxxxxxxxxxxxxxxxx			
Off-shore Wind Energy		xxxxxx	xxxxxxxxxxxx	xxxxxxxx		
Advanced Conversion		xxxxxx	xxxxxxxxxxxx	xxxxxxx		
Energy Crops		xxxxxx	xxxxxxxxxxxx	xxxxxxxx		
Tidal Power		xxxxxx	xxxxxxxxxx			
Wave Energy		xxxxxxxxxxx	xxxxxxxxxxxx			
Advanced Fuel Cells		xxxxxxxxxxx	xxxxxxxxxxxxx			
Photoconversion		xxxxxxxxxxx				
Geothermal Hot Dry Rock		xxxxxxxxxxx				
Thermal Solar	xxxxxxxx					

2.2 Prospects For The Future

World Wide

2.2.1 Recent studies suggest that renewable energy technologies could meet a growing proportion of the world's growing demands for energy at prices lower than those usually forecast for non-renewable sources. The World Energy Council (Ref 12) estimates that by 2020 the renewables contribution to total world energy supply could be almost twice the 1990 level. A United Nations (UN) report (Ref 13) indicates that by the middle of the next century renewable sources of energy could account for three-fifths of the world's electricity market and two-fifths of the market for fuels used directly. Although the UN report is less cautious than the Council in estimating what can realistically and economically be achieved in the medium term, particularly in its consideration of the contribution of modern biomass, both these and other sources indicate a potential significant increase in the world wide utilisation of new and renewable energy.

European Community

2.2.2 The European Commission indicative objective for the exploitation of renewable energy is to double the contribution renewables make to total energy consumption within the European Community. This involves an increase from the current 4% to 8% by the year 2005 with a tripling of the contribution from renewables based electricity.

United Kingdom

2.2.3 New and renewable energy sources have the potential to make a large impact on UK electricity needs in the next century, with a further significant contribution from heat sources. The Renewable Energy Advisory Group (Ref 2) considered that a plausible figure for the upper band of what is feasible by 2025 under severe pressures of need and economics would be the equivalent of around 20% of 1991 electricity supply, or about 60 TWh/y. The figure of 60 TWh/y is equivalent to about 10,000 MW DNC of renewables. The large expansion from current levels would be due to a combination of anticipated technological developments, low discount rates, a self sustaining market and increased environmental pressures on the energy industries.

2.2.4 An assessment of theoretical or 'accessible' resource for each appropriate technology at a cost of less than 10 pence per kilowatt-hour (p/kWh) is given in Figure 3. The **Accessible Resource** represents the resource which would be available for exploitation by a mature technology after only primary constraints are considered. For example, for wind power, National Parks and physical constraints such as housing, roads and lakes are excluded from the calculation of the resource size. For most technologies this measure of resource is still large and its full exploitation unlikely to be acceptable as it would result in power plant in every available location. Moreover, the cost ceiling of 10 p/kWh is high compared with the current pool price (less than 3 p/kWh on average). Whilst the Accessible Resource may therefore indicate a considerable theoretical potential for renewable energy, it is not a realistic measure of the actual contribution which may be made in future. However the Accessible Resource provides a starting point from which realistic estimates of the likely resource size can be made.

2.2.5 To assess the contribution that technologies might make in the real world a more realistic measure, taking account of additional constraints upon their deployment, is required. This measure is here called the **Maximum Practicable Resource**. In deriving

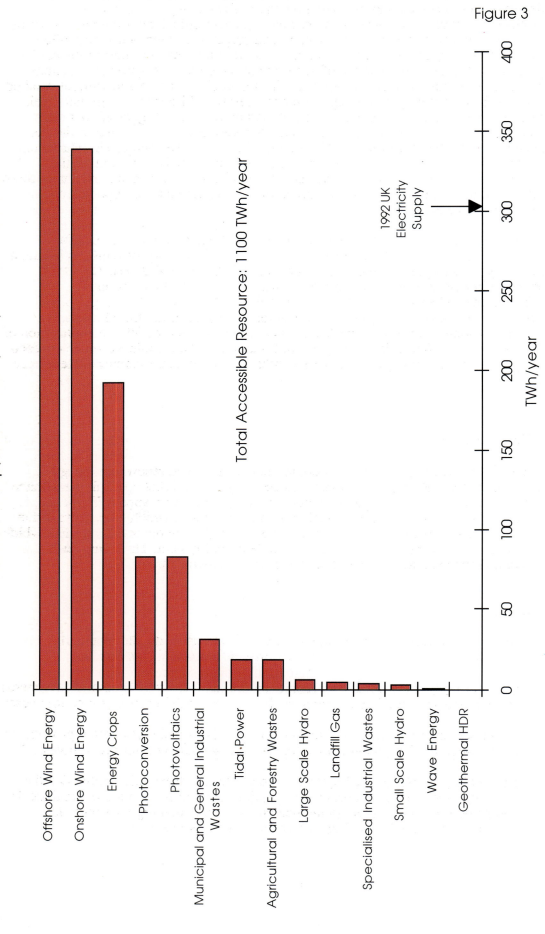

Accessible Resource for electricity producing renewable energy technologies at a cost of 10p/kWh or less (1992) 8% discount rate

Total Accessible Resource: 1100 TWh/year

1992 UK Electricity Supply

TWh/year

Offshore Wind Energy
Onshore Wind Energy
Energy Crops
Photoconversion
Photovoltaics
Municipal and General Industrial Wastes
Tidal Power
Agricultural and Forestry Wastes
Large Scale Hydro
Landfill Gas
Specialised Industrial Wastes
Small Scale Hydro
Wave Energy
Geothermal HDR

0   50   100   150   200   250   300   350   400

Figure 3

Photovoltaics and Photoconversion will compete for sites as will Crops and Onshore Wind.

Source: reference 3

estimates of the Maximum Practicable Resource an examination of the constraints on the deployment of each technology, and how they might change with time, was undertaken. Many of these constraints - regulatory, sociological, environmental - are not susceptible to objective scientific assessment and subjective judgements were often required. These judgements were informed by studies undertaken as part of the earlier programme and existing experience of deployment which in most cases is extremely limited. Moreover for each technology the UK as a whole was assessed without detailed analyses of the complex system integration and operational issues at a regional level, but ongoing regional assessments (Refs 18-20 and others) have suggested that these issues could significantly constrain the Practicable Resource. The resource estimates presented here are therefore for the "Maximum" Practicable Resource. The Maximum Practicable Resource is expressed in terms of cumulative supply curves in Figures 4-7 for the years 2005 and 2025 at both 8% and 15% discount rates. When considering these curves it is important to refer to the qualifications attached to each component cost curve in Reference 3.

2.2.6 Estimates of the contributions renewables could make to the year 2025 in competition with other supply options were made using an energy systems model and a range of scenarios as part of an overall appraisal of UK energy RDD&D (Ref 10) summarised in Annex 3. As an illustration it was estimated that the economic contribution from electricity producing renewables in 2025 lies in the range 15-190 TWh/y. The cumulative supply curves of Figures 4-7 representing the Maximum Practicable Resources for the technologies, were inputs to that model. An explanation of the derivation of the Accessible Resource and Maximum Practicable Resource for each of the appropriate technologies is provided within the individual modules of Reference 3.

2.3 Constraints

2.3.1 Until very recently, the minimal deployment of new and renewable energy technologies in the UK meant that there was essentially no market operating and hence limited interest on the part of suppliers, potential developers or planners. These technologies are unlikely to be exploited under the conditions which have prevailed in the UK without stimulation by Government and removal of inappropriate barriers which would otherwise prevent or at least slow down their deployment. These market imperfections span technical, institutional and environmental areas:

Technical

2.3.2 Many of the technologies are immature and need further development and demonstration before they can become competitive. Particular issues are:

-   **Cost reduction:** technical development and establishment of production systems and markets on a sufficiently large scale are needed to improve competitiveness;

-   **Risk:** demonstration and establishment of standards and certification procedures are required to give potential developers the necessary confidence to invest;

-   **Integration of distributed systems** into electricity distribution networks: the planning, operational and protection requirements for integration of small-scale and sometimes intermittent plant, need to be understood together with the requirements for additional transmission lines to remote areas.

Figure 4

Supply curves for electricity producing renewable energy technologies:
Maximum Practicable Resource, 2005, 8% discount rate

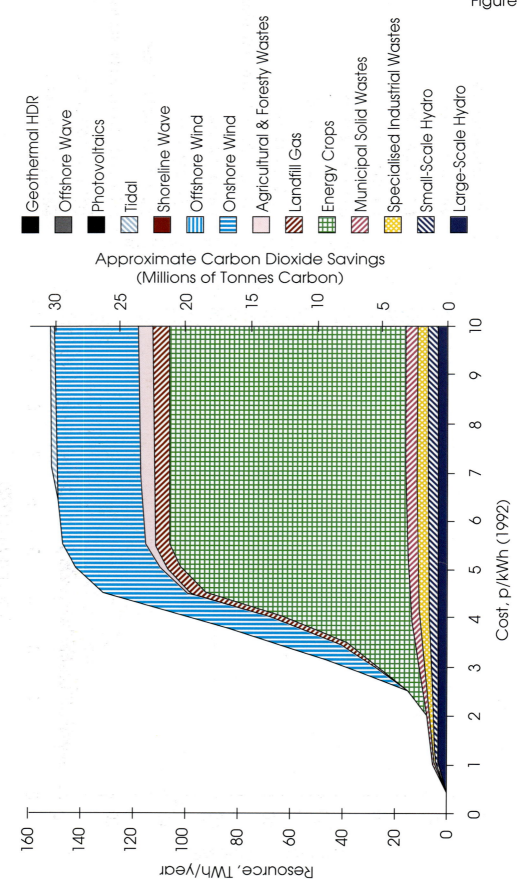

Source: reference 3

Figure 5

Supply curves for electricity producing renewable energy technologies:
Maximum Practicable Resource, 2005, 15% discount rate

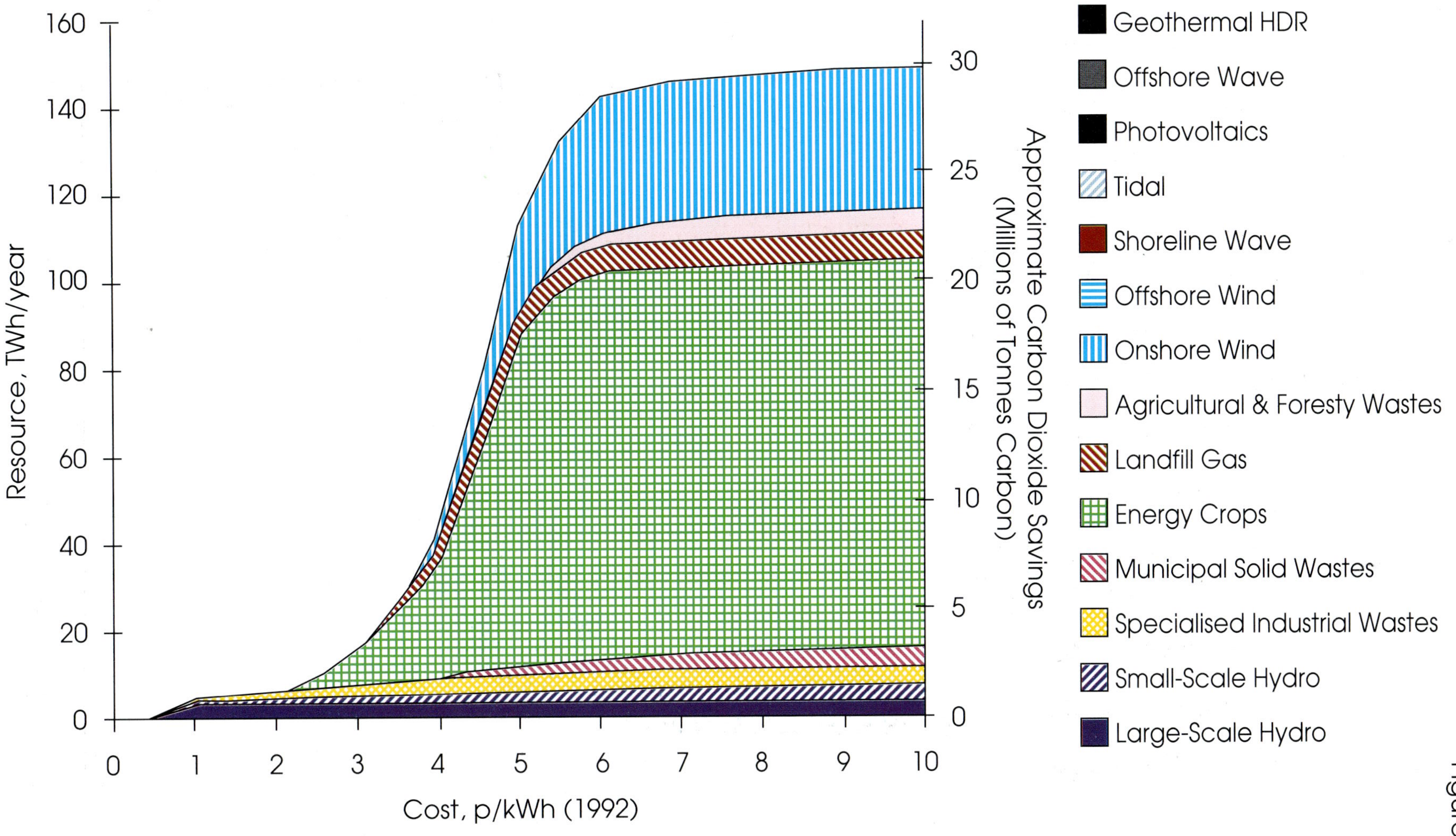

Source: reference 3

Figure 6

Supply curves for electricity producing renewable energy technologies:
Maximum Practicable Resource, 2025, 8% discount rate

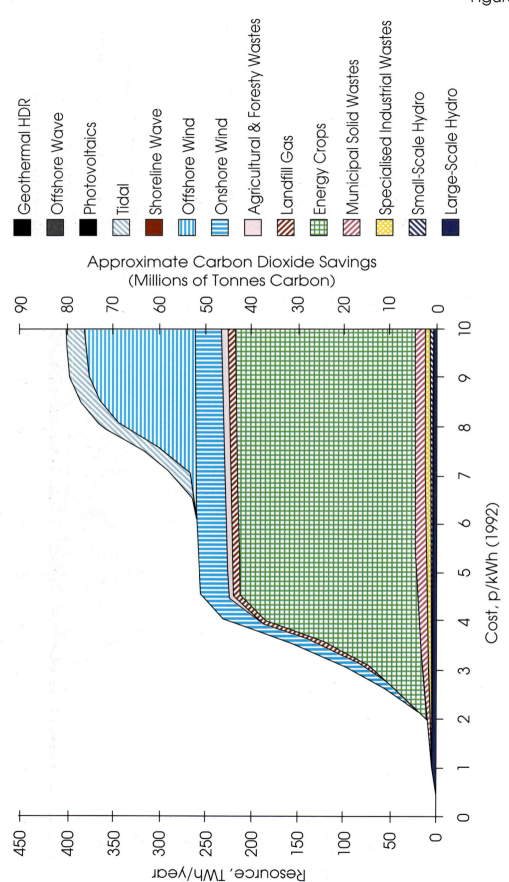

Source: reference 3

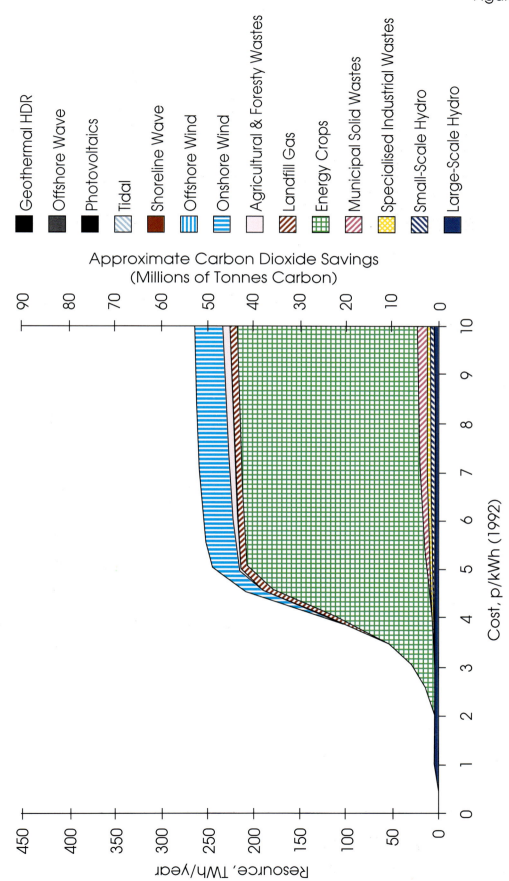

Supply curves for electricity producing renewable energy technologies:
Maximum Practicable Resource, 2025, 15% discount rate

Figure 7

Legend:
- Geothermal HDR
- Offshore Wave
- Photovoltaics
- Tidal
- Shoreline Wave
- Offshore Wind
- Onshore Wind
- Agricultural & Foresty Wastes
- Landfill Gas
- Energy Crops
- Municipal Solid Wastes
- Specialised Industrial Wastes
- Small-Scale Hydro
- Large-Scale Hydro

Approximate Carbon Dioxide Savings
(Millions of Tonnes Carbon)

Resource, TWh/year

Cost, p/kWh (1992)

Source: reference 3

13

Institutional

2.3.3 Legal, administrative and financing systems have all evolved in terms of conventional mainly large-scale energy supply plant whereas new and renewable energy technologies tend to be small-scale, often located in rural areas and unfamiliar: all of which impede their development. Moreover the development companies involved are usually small and medium sized enterprises (SMEs). Some of the issues are:

- **Establishment of an Industries:** a market for heat and electricity from new and renewable energy sources needs to be established in order for industrial opportunities, largely for SMEs, to be realised;

- **Awareness:** there is a requirement for appropriate information, promotion of innovation, spread of best practice and the transfer of technology in order to secure industrial competitiveness. There is also a need to inform public opinion, political processes and commercial and industrial decision making, of new and renewable energy;

- **Skills:** education and training are necessary to provide the skills base for rapidly developing industries;

- **Finance:** due to the perceived risks associated with the deployment of new energy technologies investors currently demand high rates of return. Methods by which new energy technologies can be financed are required to facilitate their deployment particularly by SMEs;

- **Planning procedures:** the planning process needs to be properly informed of the impact of the various technologies, so that procedures can be developed which will allow appropriate deployment to take place.

Environmental

2.3.4 All energy technologies have environmental impacts. Although the majority of the public appear, from surveys conducted, to be strongly supportive of renewable energy, there are inevitably some objections to the siting of plant. Environmental constraints include the following:

- **The delivered cost of energy** reflects compliance with environmental performance standards but not necessarily the full cost of environmental impact or externalities. Accounting for the cost of environmental externalities associated with all energy sources would be required to facilitate comprehensive appraisal of competing energy supply options;

- **Planning**: an equitable balance needs to be struck between the regional and global benefits of new and renewable technologies in reducing harmful emissions, and localised environmental concerns.

14

2.4 Benefits

2.4.1 New and renewable energy sources can potentially contribute to energy needs in a significant and sustainable way, helping to reduce emissions of pollutant gases. They can also increase the regional and national diversity, and overall security, of supply. In the long term this may reduce the cost of energy by increasing the number of options available, especially those which could be used for coping with environmental threats. There is also the prospect of industries growing to compete in expanding export and domestic markets and much of this new industry could be expected to be located in rural areas.

2.4.2 New and renewable energy sources are an important element, particularly in the medium to long term, of a broad based approach by Government to reduce environmental damage caused by emissions of polluting gases. This approach was outlined in the UK Programme on Climate Change (Ref 14) and the UK Strategy for Sustainable Development (Ref 15). Increasing use of new and renewable energy will help to reduce environmental damage from acid rain, and to fulfil the commitment, under the United Nations Framework Convention on Climate Change, to take measures aimed at returning emissions of carbon dioxide and other greenhouse gases to 1990 levels by the year 2000. The increase in the figure for new renewable electricity generation capacity announced in the Coal Review (Ref 5) should contribute about an additional 0.5 million tonnes of carbon dioxide (as carbon) annually towards this by 2000. Additionally, support for new and renewables assists in meeting the Government's undertakings towards sustainable development with the adoption of Agenda 21 at the United Nations Conference on Environment and Development in June 1992. The UK new and renewable energy programme also endeavours to contribute to the European Commission's indicative objectives for reducing carbon dioxide emissions by developing relevant technologies.

2.4.3 There is only an approximate equivalence between the amount of renewable energy used to displace fossil fuel based electricity generation plant, and savings in emissions of carbon dioxide (refer Annex 2). Figures 4 to 7 therefore include scales indicating the approximate range of savings in emissions of carbon dioxide, based on displacement of the 1990 typical mix of fuels. Table 1 presents estimated ranges of the economic potential of electricity producing renewables for 2005 and 2025, with the approximate equivalent carbon dioxide savings.

Table 1: Economic Potential of Electricity Producing Renewables in the UK			
	1992	2005 (note 1)	2025
Electricity supplied by renewables (TWh/y) (% of 1992 UK electricity)	6 2%	15-75 5%-25%	15-190 (note 3) 5%-63%
Emissions of carbon dioxide saved (mtC/y) (note 2) (% of 1992 UK emissions)	1 1%	3-15 2%-9%	3-38 2%-24%

Note 1: The range for 2005 includes contributions from existing plant and equipment.
Note 2: millions of tonnes of carbon dioxide (as carbon) annually, based on the 1990 typical mix of fuels.
Note 3: REAG considered a range of 15-60 TWh/y to be more plausible

2.4.4 The range of contributions in 2025 of 15-60 TWh/y, considered plausible by REAG, would be equivalent to a range of savings in greenhouse gas emissions of approximately 3 to 12 million tonnes of carbon (10 and 40 million tonnes of carbon dioxide) annually. By 2000, the figure of 1,500 MW DNC (equivalent to about 9 TWh/y of electricity) could contribute possibly 2 million tonnes of carbon dioxide savings annually (as carbon), dependent upon the type of displaced fuel and the mix of renewable technologies deployed. In addition, this level of renewables based electricity production would have a substantial impact on emissions of other pollutants, notably sulphur dioxide (SO_2) and oxides of nitrogen.

2.4.5 The knowledge, equipment, and systems required to use new and renewables have a very significant industrial potential for both home and export markets. The European Commission ALTENER proposal of 1992 indicates that the size of the annual world-wide market for renewables production equipment had been estimated to be about ECU 40 billion. In the UK if the figure of 1,500 MW of new renewable electricity generating capacity by 2000 were achieved, it would involve sales of equipment and systems to a value of about £3,000 million. Export sales would be in addition to this total, and it has been estimated that the UK photovoltaic industry alone could generate exports amounting to some £700 million annually. Another example of a technology with export potential, is wind. By 1993 there were more than 3,000 MW (rated) of wind energy capacity installed throughout the world. By the year 2000, this should increase to 8,000 MW, if national targets are met. The cost of these additional installations will be around £5,000 million. British manufacturers would have an opportunity to enter such a market.

3. PROGRAMME STRATEGY

3.1 Strategy

3.1.1 The Government has initiated a market enablement strategy to implement its policy, stimulating the development of sources and industrial and market infrastructure so that new and renewables are given the opportunity to compete equitably with other energy technologies in a self sustaining market. In so doing, the strategy aims to help the technologies become economic so that they can compete without support in the longer term. In the shorter term, the programme seeks to minimise the costs of the technologies and the overall support provided by the energy consumer.

3.1.2 This strategy involves:

-   stimulating an initial market for electricity producing technologies close to commercial competitiveness via the NFFO;

-   assessing and developing technology options;

-   ensuring that the market is fully informed;

-   removing inappropriate market barriers;

-   encouraging internationally competitive industries to develop.

16

3.1.3 Each new technology requires assessment of its potential and eventual commercial prospects in home and export markets. For those shown to have prospects research, development and field trials in collaboration with suppliers and developers are necessary. Where the technology appears to be technically viable and has commercial prospects an initial, and probably protected, market needs to be stimulated to allow it to become competitive and demonstrate a degree of commercial maturity. If successful, that technology will then compete if the institutional and commercial framework is equitable. Unaided commercial deployment will follow, the extent depending upon normal market mechanisms.

3.1.4 Each stage of this process requires increasing levels of investment. It is appropriate for Government to be heavily involved in the early stages, withdrawing as the framework evolves and commercial competitiveness approaches. The DTI role over this period is as a "task force" stimulating the other industrial, commercial, Local Authority and Government organisations to play their role in enabling the exploitation of new and renewable energy technologies where appropriate.

3.1.5 The strategy follows this approach, focusing primarily upon technologies with most promise based upon a consideration of all the information prepared during the course of the strategic review, including the assessment of their potential economic contribution in the UK and the timing of that contribution derived from the scenario analysis outlined in Annex 3. On this basis the new and renewable energy technologies have been classified as presented in Table 2. Technologies identified as likely to be contributing economically to the UK electricity supply by the year 2005 will be given market support via the NFFO. These include wind, hydro, landfill gas, municipal and industrial waste, energy crops and agricultural and forestry wastes. Appropriate RDD&D support will also be provided, subject to overall Government expenditure pressure. Technologies at an early stage of assessment will be investigated to provide sufficient information to form a reliable view of their potential, and if shown to have sufficient prospects RDD&D support may be provided. This group includes photovoltaics, fuel cells, active solar and photoconversion. Technologies such as wave, geothermal and tidal, which are unlikely to contribute significantly by the year 2025, will be subject only to a watching brief. Annex 1 outlines the overall work programme and Annex 2 contains the aims and description for each of the component new and renewable energy technology sub-programmes.

Non-Fossil Fuel Obligation

3.1.6 A purpose of the renewables NFFO (see Box 1) is to demonstrate those electricity producing technologies which are closest to becoming commercially competitive and thereby facilitate their entering the market where, once established, it is expected they will become viable without further special support. The making of NFFO Orders is the Government's main instrument for working towards the figure of 1,500 MW DNC of new renewable electricity generating capacity in the UK by 2000. A possible projection of UK NFFO capacity, split by technology, is presented in Figure 8. These indicative trends will be dependent upon many factors, but an underlying principle is the requirement for steady convergence under successive Orders between the price paid for electricity under the NFFO and the market price. This will only be achieved if there is effective competition in the allocation of NFFO contracts. By 1993, NFFO was supporting renewables to the extent of some £30 million annually, and this is likely to build up over the next few years to a maximum of about £150 million annually, financed through a levy of about 1% on electricity prices, before reducing.

Potential NFFO within the Programme - Illustrative Split by Technology

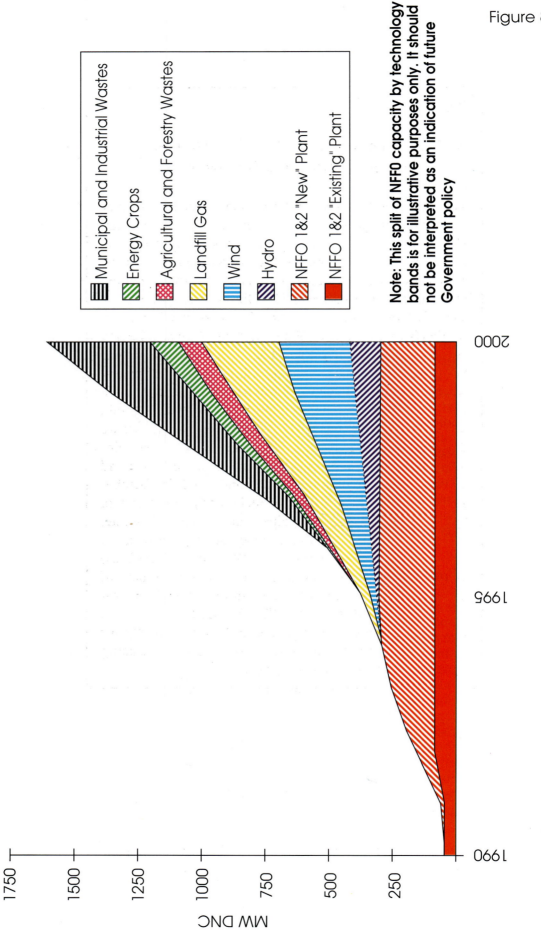

Figure 8

18

Box 1

NFFO - a Guaranteed Premium Market for Electricity from Renewables

The Electricity Act (1989) and broadly similar legislation in Northern Ireland empowers the relevant UK Secretary of State to make Orders requiring the Public Electricity Suppliers (PESs) to secure specified amounts of electricity generation capacity from specified renewable energy sources. The Obligations arising from such Orders are known as Non-Fossil Fuel Obligations (NFFOs) in England & Wales and in Northern Ireland and as the Scottish Renewables Obligation (SRO) in Scotland. The PESs meet those Obligations by contracting with renewables-based generators; and the relevant Office of Electricity Regulation (Great Britain or Northern Ireland) vets those arrangements to ensure that they will secure the capacity specified. For the third round of bidding in England & Wales and the first round in Scotland, the intention was announced in October 1993 to make Orders for five technology bands (wind, hydro, landfill gas, municipal and industrial waste, energy crops and agricultural annd forestry wastes), for overall capacities of 300 - 400 MW (England &Wales) and 30 - 40 MW (Scotland). Decisions on the size of bands are expected in late 1994 - in the light of the cost and quality of bids received. The first Order in Northern Ireland (NI-NFFO-1) is expected to be made shortly for up to 15 MW, although no specific banding arrangements have been announced.

Table 2: Classification of Technologies

Market enablement via NFFO and/or RDD&D:

- Passive Solar Design
- Agricultural and Forestry Wastes
- Municipal and Industrial Wastes
- Advanced Conversion Technologies
- Hydro Power (small, new)
- Landfill Gas
- Wind Power (on-shore)
- Energy Crops

Assessment, RDD&D:

- Photovoltaics
- Photoconversion
- Advanced Fuel Cells
- Active Solar

Watching Brief:

- Wave Energy
- Geothermal Hot Dry Rock (HDR)
- Geothermal Aquifers
- Wind Power (off-shore)
- Tidal Power
- Hydro Power (large, new)
- Ocean Thermal Energy Conversion
- Thermal Solar and other technologies not listed

The Supporting RDD&D Programme

3.1.7 The strategy envisages an RDD&D programme extending to the year 2005, but with a major review after five years. This programme, complementary to NFFO, provides an assessment of the technology options and potential; stimulation of technology development and cost reduction with industry; monitoring of developments under NFFO; removal of inappropriate market barriers; and technology transfer facilities. The programme will commission work and respond to the needs of industry, as well as sponsor and promote the new and renewable energy technologies within other UK and international programmes. It is being undertaken in collaboration with business (users and suppliers), planners and financiers.

3.1.8 The programme will build upon the foundation laid by the work to date. There will continue to be a strong focus on competitiveness and wealth creation, with promotion of innovation, best practice and the transfer of technology. The programme will continue helping firms gain access to science and technology, irrespective of its source, be it national or international. It will continue to facilitate appropriate deployment of the technologies, and to help build innovation partnerships between firms, academe and overseas organisations. There will be continuation of support for business, academic, and industrial groups serving the home and export markets. This support is provided from within the new and renewable energy technology programme, and complemented by other Government mechanisms such as LINK, EUREKA, SMART, SPUR, and DOE's Best Practice programme.

3.1.9 A major aspect of the programme is international collaboration, especially integration with the European Community's activities. It is recognised that the most efficient use of resources, especially in undertaking pre-competitive research and development, is often to participate within international frameworks. Most OECD* nations support substantial programmes of RDD&D for renewable energy technologies and many have assistance schemes to facilitate deployment of renewable technologies into the energy market place. Valuable groupings are available in particular under the aegis of the International Energy Agency (IEA) and European Community, and this receives a high priority within the work programme. The programme also facilitates the involvement of organisations in support of international aid programmes.

3.2 Expenditure

3.2.1 Estimates of the funding of UK RDD&D for new and renewable energy technologies, including contributions from industry and others, are presented in Figure 9a. This Figure describes the historic Government support and includes a broad estimate of the future funding by the DTI. However the resources allocated to each sub-programme, and for the total programme, will be determined at the appropriate time as part of the Government expenditure cycle. Moreover all DTI expenditure, including energy R&D, is subject to review and the strategy may be amended in the light of that, and of the availability of resources. The duration of the programme recognises the anticipated time scales for developments in the cost and performance of the various technologies. Although the programme is therefore presented as extending to 2005, there is a commitment to undertake a fundamental review within five years, and continuation of the work would be dependent upon the outcome of that exercise.

3.2.2 The RDD&D programme budget for 1994/95 is £19.78 million and is expected to reduce thereafter over a ten year horizon as the technologies move towards the market and the DTI's "task force" role diminishes. The main vehicle for financial support during the market stimulation period will be the NFFO and subsequently unaided capital investment in plant will follow if the technologies are competitive.
This evolution of the market for the technologies is depicted in Figure 9b, which is an estimate of overall investment within the UK on new and renewable energy. The Figure illustrates the instrumental role being undertaken by the Government's programme as the technologies mature from research through to commercial deployment. It is anticipated that the resultant capital investment by industry could exceed £3,000 million within the time frame being considered. Because of the uncertainty in predicting the future energy market, the estimate of investment in commercial plant can only be a broad approximation.

*Organisation for Economic Co-operation and Development

Figure 9a

Funding of RDD&D for New and Renewable Energy Technologies

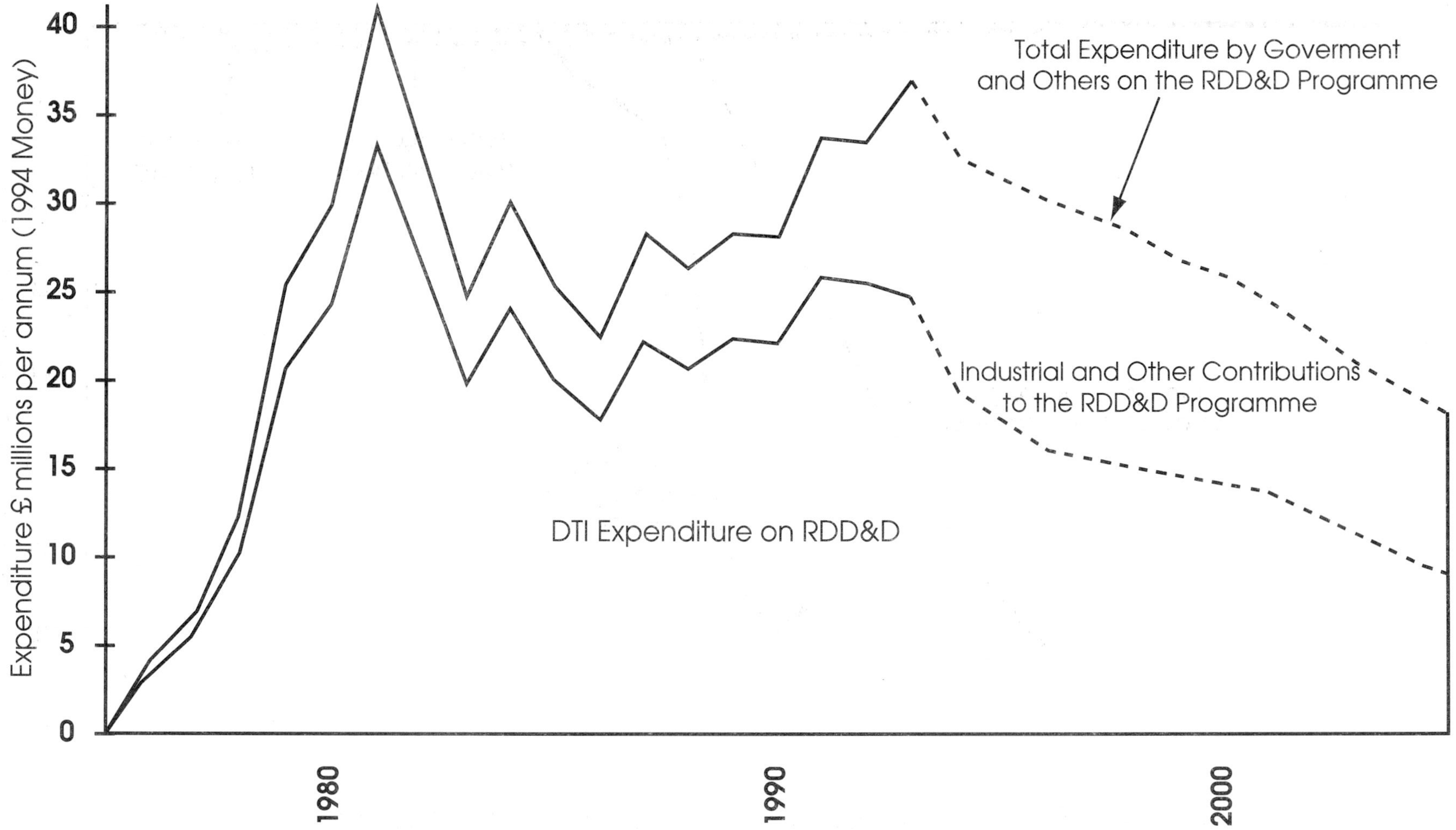

Total Expenditure by Goverment and Others on the RDD&D Programme

Industrial and Other Contributions to the RDD&D Programme

DTI Expenditure on RDD&D

Expenditure £ millions per annum (1994 Money)

40   35   30   25   20   15   10   5   0

1980   1990   2000

Future costs can only be rough estimates at present, both in total and the split between Government and Industry. The resources needed will have to be decided at the appropiate time as part of the Goverment expenditure cycle

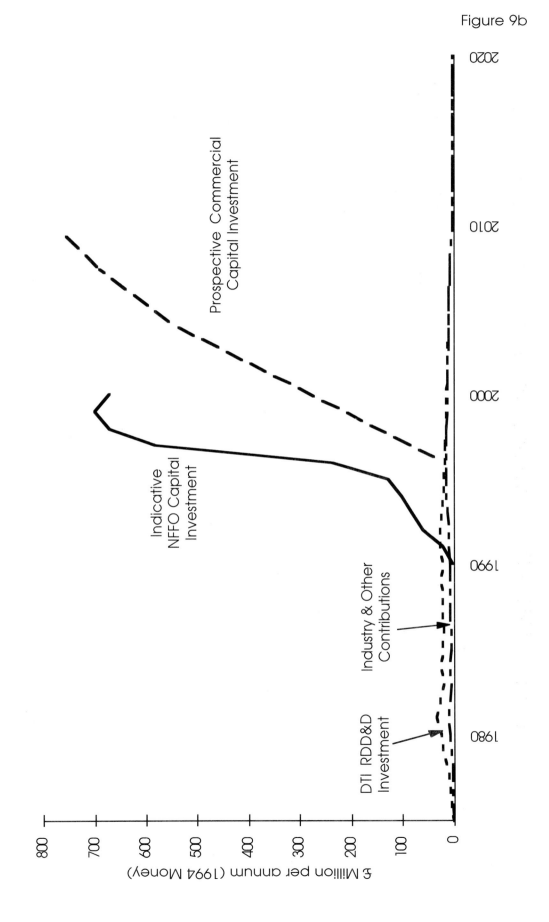

Potential Investment in New and Renewable Energy Technologies

Figure 9b

23

REFERENCES

1. Energy Select Committee: Fourth Report on Renewable Energy, Volume 1. March 1992, HMSO.

2. Renewable Energy Advisory Group: Report to the President of the Board of Trade, Energy Paper Number 60. November 1992, HMSO.

3. An Assessment of Renewable Energy for the UK, ETSU-R82. 1994, HMSO.

4. Review of Wave Energy, ETSU-R72. December 1992.

5. The Prospects for Coal: Conclusions of the Government's Coal Review Cm 2235. March 1993, HMSO.

6. ALTENER Programme: Council Decision of 13 September 1993 Concerning the Position of Renewable Energy Sources in the Community (93/500/EEC).

7. This Common Inheritance: Britain's Environmental Strategy Cm 1200. September 1990, HMSO.

8. Realising our Potential: A Strategy for Science, Engineering and Technology Cm 2250. May 1993, HMSO.

9. Renewable Energy in the UK: The Way Forward. Energy Paper Number 55, June 1988. HMSO.

10. Energy Technologies for the UK: An Appraisal of UK Energy RDD&D, ETSU. This document is expected to be published in Spring 1994 as Energy Paper Number 61, HMSO.

11. IEA/OECD, Energy Statistics and Balances of Non-OECD Countries 1990-1991. 1993.

12. World Energy Council: Renewable Energy Resources - Opportunities and Constraints 1990-2020. 1993.

13. Renewable Energy, ed Johansson et al. 1993. This document was commissioned by the UN Solar Energy Group for Environment and Development: a Group convened under mandate of the General Assembly Resolution A/45/208.

14. Climate Change: The UK Programme Cm 2427. January 1994, HMSO.

15. Sustainable Development: The UK Strategy Cm 2426. January 1994, HMSO.

16. Digest of UK Energy Statistics. 1993, HMSO.

17. National Audit Office report on the Renewable Energy Research, Development and Demonstration Programme. January 1994, HMSO.

18. Prospects for Renewable Energy in Northern Ireland. NIE, ETSU, DED. July 1993.

19. An Assessment of the Potential Renewable Energy Resource in Scotland. Scottish Hydro Electric, Scottish Power et.al. December 1993.

20. Renewable Sources of Electricity in the SWEB Area: Future Prospects. ETSU, SWEB. June 1993.

ANNEXES

THE WORK PROGRAMME

1. This Annex presents the DTI's work programme that has been established to meet the Government's aims and objectives for new and renewable energy sources (see below). A wide range of integrated activities from legislative measures through to the support of R&D on novel technologies is required. Many of the programme activities, particularly those which deal with the institutional and legal framework, require action to be undertaken directly by the Department of Trade and Industry. Other activities, particularly those which relate to the development, commercialisation and marketing of technologies, including the management of external contracts, are met by actions undertaken within specific sub-programmes as detailed in Annex 2.

Programme Aims and Objectives

- Encouraging the uptake of the technologies which are approaching competitiveness by:
    - assessing when technologies become cost effective;

    - establishing an initial market via the NFFO mechanism;

    - removing inappropriate legislative and administrative barriers;

    - ensuring that the market is fully informed;

    - stimulating development of the technologies as appropriate.

- Assessing technologies which have prospects for the longer term, and maintaining the option of developing and deploying at a later stage.

- Encouraging internationally competitive industries to develop and utilise capabilities for the domestic and export markets, taking account of those factors which influence business competitiveness.

- Quantifying environmental improvements and disbenefits associated with the new and renewable energy technologies.

- Managing the programme effectively.

Key Themes of the Work Programme

2. The RDD&D programmes which were initiated in the UK and elsewhere in the 1970s have brought many of the most promising new and renewable energy technologies close to the point of commercial application. This success has resulted in a change of emphasis within the UK programme. Today, the most important task is to bring the near commercial technologies into the market and many of the activities in the new programme are concerned with market enablement and stimulation. The most important mechanism for market enablement is the NFFO, and progress to end 1993 is presented in Table 3.

Table 3: NFFO Projects Operational at end 1993

Technology	1990 Order (NFFO-1)		1991 Order (NFFO-2)	
	No of Projects Generating	Contracted Capacity MW, DNC	No of Projects Generating	Contracted Capacity MW, DNC
Hydro	20	9	7	10
Landfill Gas	19	30	25	44
Municipal Solid Waste Incineration	4	40	1	7
Other Waste Incineration	3	25	0	0
Sewage Gas	7	6	19	27
Wind	8	12	21	41
Totals	**61**	**122**	**73**	**129**

3. To allow the smooth introduction of new and renewables into the market, while NFFO mechanisms are available and thereafter, the programme attempts to ease inappropriate institutional constraints by addressing such matters as market awareness, planning procedures, financing arrangements, and the shortage of skilled personnel. The nature of the supporting RDD&D has also changed as the technologies have approached the market. Whereas in the past emphasis was upon the technologies, today emphasis is upon assessing potential markets and developing and adapting products to meet market requirements. For example, the programme addresses such matters as: the detailed assessment of the location and exploitation potential of the renewable resource; the development of less intrusive technologies; assessment of environmental impacts; and system planning and integration for intermittent energy sources.

4. In summary, the scope of the activities undertaken by the programme include:

- stimulating an initial market for electricity producing technologies close to commercial competitiveness via the NFFO;

- assessing and developing technology options;

- ensuring that the market is fully informed;

- removing inappropriate market barriers;

- encouraging internationally competitive industries to develop.

Key Activities

5. Under the above headings, the key activities being undertaken include the following:

Stimulating an initial market

- To make progress towards the 1,500 MW DNC figure by 2000 (equivalent to about 9 TWh/y of electricity).

- To reduce the premiums paid to particular technologies under successive Orders.

- To provide advice on the arrangements for each of the future NFFO Orders.

- To assess the need, and possible mechanisms, for an initial market support scheme for heat producing new and renewable energy technologies.

Assessing and developing technology options

- To stimulate the development of technologies, undertaking RDD&D and introducing each technology into NFFO as appropriate.

- To assess each of the technologies approaching competitiveness to determine when they would become cost effective.

- To assess technologies with prospects of becoming longer term options undertaking R&D if necessary otherwise maintaining a watching brief.

- To quantify environmental improvements and disbenefits associated with new and renewable energy technologies.

- To assess the contribution that new and renewable energy technologies might make to meeting national and international environmental targets.

- To review regularly the programme as a whole.

- To review regularly and publish the prospects for new and renewable energy technologies in the UK and the progress in achieving the aims of the Government's programme.

- To achieve maximum benefit from collaboration, recognising the contribution that other programmes (for example EC, IEA, LINK, SMART, SPUR and EUREKA) can make.

Ensuring that the market is fully informed

- To conduct business development strategies with utilities and major developers.

- To understand key industrial and commercial market sectors and to develop advisory and training support for them, enabling early development of new and renewable energy technologies.

- To develop educational resources.

- To provide timely, accurate, comprehensive and independent information to all sectors.

Removing inappropriate market barriers

- To establish an appropriate financing regime for new and renewable energy technologies.

- To liaise with DOE on the provision of planning advice.

- To ensure that where appropriate, coverage of technical standards and certification procedures for new and renewable energy technologies is comprehensive and applicable throughout the European Community.

Encouraging internationally competitive industries to develop

- To produce an assessment for the export prospects of new and renewable energy technologies.

- To conduct business development studies with technology suppliers.

- To establish a system to inform industry of export prospects for new and renewable energy technologies.

The Sub-Programmes

6. Most of the sub-programmes are technology specific, addressing various aspects of a particular technology's market enablement needs. The role of these sub-programmes is to assess and develop promising technologies and then to introduce them to the market. As the technologies vary in their nature and intended market and are at different states of development, their requirements differ. The development that a typical technology undergoes from its conception to market deployment, and the type of activities undertaken at each of these stages, are as follows:

Development Stage	Activities which may be supported by a typical technology development sub-programme
Conception	Preliminary technical feasibility assessment.
Research	Initial resource and market assessment; Initial assessment of economics; Fundamental research; Environmental assessments.
Development	Design studies; Equipment development; Prototype testing; Assessment of resource, economics and environmental impacts.
Demonstrate	Pilot demonstration; Full scale demonstration; Independent monitoring; Assessment of resource, economics and environmental impacts.
Deployment	System integration considerations; Environmental impact monitoring; Assessment of resource, economics and environmental impacts.
Dissemination	Planning studies and guidance; Best Practice case studies; Market research; Market awareness - publications, exhibitions, seminars, etc.; Advice and training.

7. To assist and complement the work of these technology specific sub-programmes, two cross technology sub-programmes, Commercialisation and Marketing, have been established.

8. The Commercialisation sub-programme was created to address the cross-technology issues and non-technical barriers that could impede the commercial deployment of new and renewable energy technologies. Many of these barriers result from the lack of experience with the technologies of those involved with deployment - planners, financiers, lawyers, regulators and even developers themselves. The Commercialisation sub-programme works with the technology development sub-programmes to identify and tackle the generic non-technical barriers, provides expertise in these areas and co-ordinates the relevant activities across the DTI new and renewable energy programme.

9. The purpose of the Marketing sub-programme is to disseminate information generated by the individual technology sub-programmes, the Commercialisation sub-programme and the NFFO mechanisms in a targeted and cost-effective manner, in order to secure wider deployment of the new and renewable energy technologies. Underpinning this is market research to increase market understanding and knowledge, and to provide feedback on the impact and effectiveness of specific marketing and promotion activities to the other sub-programmes.

10. A description of the sub-programmes is presented in Annex 2.

Programme Management, Co-ordination, Evaluation and Reporting

11. The aims of the new and renewable energy programme demand a wide range of actions and activities to be undertaken and call for a range of management skills much wider than those normally associated with technology development programmes. The success of the programme will be determined largely by the degree to which these separate elements are integrated into a cohesive whole. The approach taken to the management and execution of the programme is, therefore, crucial.

12. An essential part of the programme is the review of strategy after 5 years and the regular reviews of progress with the individual programme elements. A review of progress with each of the sub-programmes and an update of the prospects for the technologies will continue to be carried out annually using the existing structure of advisory committees both at the individual sub-programme level and at the overall programme level. The annual review of the programme involves scrutiny by the DTI's Renewable Energy Advisory Committee (REAC) which includes representatives from industry, commerce and academe. A summary of the information produced for that committee will continue to be published annually. Liaison and co-ordination with other Government departments and international programmes are important and will continue.

13. Evaluation of the programme is carried out in accordance with the DTI's formal evaluation procedures. The information necessary for this evaluation is gathered on a continuous basis as part of the management information procedures associated with the programme. The effectiveness of the various measures and activities carried out under the programme are evaluated over appropriate time scales and value for money assessed.

TECHNOLOGY DESCRIPTIONS

AND SUB-PROGRAMMES

Contents

Wind Energy

The Technology

Description and Present Status

Technological advances have produced wind turbine generators which efficiently convert wind power into rotating shaft power and then into electricity. There are two basic design configurations - horizontal axis machines and vertical axis machines. Horizontal axis designs are at a more advanced stage of development and the evidence is that they are also more cost-effective. Vertical axis designs may offer scope for larger offshore developments. Apart from the need to demonstrate adequate lifetimes, there is no doubt about the technical feasibility of harnessing wind power, especially for land-based sites.

The existing technology offers a range of power ratings from a few kilowatts up to several megawatts. The technology is well established, with over 20,000 grid connected machines in operation world-wide. Current development work is concentrating on reliability, the further reduction of cost and noise emissions, aspects of the electrical generation and overall performance.

Wind turbines are available as "off the shelf" equipment and a wind farm can be installed and operating within about a year from its conception. Moreover, it is simple to decommission wind generator installations at the end of their lifetime.

In the United Kingdom, the potential of wind energy for the large-scale generation of electricity has been under systematic investigation for over 15 years through programmes involving the Department of Energy (later the Department of Trade and Industry), the Electricity Supply Industry and the manufacturers. The aim of the programmes has been to take the development of wind energy technology for large-scale electricity production through the demonstration phase and on towards commercialisation.

To date the programme strategy of the DTI (DEn) has involved:

- assessment of the technology through the construction, operation and monitoring of prototype wind turbines;

- extensive underlying R&D programme;

- investigation of the potential for reduction in the cost of energy from larger machines;

- support of selected wind farms under the NFFO to obtain data on the environmental, technical and economic aspects of commercially available machines and to obtain system operating experience.

In addition the DTI has an ongoing programme to assess the available resource and the likely cost of exploiting it.

DTI/DEn Programme Funding to 31/3/93	£54.1M
Number of R&D Projects to 31/12/93	315
Number of Commissioned NFFO Projects MW DNC to 31/12/93	29 53

Table 1: Estimated programme spend (money of the day) and number of projects

Resource

The UK has the best wind resource in Europe. There are limitations on the availability of land for wind turbine sites due both to physical constraints - such as the presence of towns, villages, lakes, rivers, woods, roads and railways - and institutional constraints such as the protection of land areas designated as being of national importance. Also, wind turbines have to be located at some distance from habitation for environmental reasons. Offshore there is potentially a very large wind resource but it will require additional technology development before it can be effectively exploited.

	Accessible Resource (TWh/year) at less than 10p/kWh (1992), 8% Discount Rate.	Maximum Practicable Resource (TWh/year) in 2005 at less than 10p/kWh (1992), 8% Discount Rate.
Onshore Wind	340	55*
Offshore Wind	380	0

Table 2: UK Wind Energy Resource (see ref 3).
* assuming no limitation imposed by integrating into the grid.

Environmental Aspects

Wind offers a clean source of electrical energy, with no production of particulates, NO_x, SO_x or CO_2. Wind turbines appear to have only a minimal effect on animal and bird life and only 2% of the land over which a windfarm extends is unavailable for other purposes, so the rest can still be used for agricultural and recreational activities.

Wind installations can be visually intrusive and emit some noise and electromagnetic interference. Careful siting is the key to minimising these impacts, avoiding nuisance and gaining public acceptance.

Economic Prospects

The cost of electricity from wind turbines is site-specific, depending on factors such as the mean wind speed and the distance from the nearest grid connection. Various components of the overall cost, such as capital equipment, installation and annual running costs are now better known as the technology is becoming established world-wide. The main uncertainty is long term reliability. The current cost of onshore wind turbines is about £800-1200/kW installed, depending on location and design. A typical onshore wind turbine would generate electricity at a cost of about 7p/kWh assuming a 20 year life, 7.5m/s wind speed at hub height and a 15% required rate of return. Electricity from offshore wind turbines is estimated to be around twice as expensive as that from onshore machines.

Opportunities and Constraints

Opportunities

- Wind energy has the potential to make a significant contribution to the UK electricity supply and helps to diversify the sources of supply.

- Wind energy offers business opportunities for UK industry at home and abroad.

- Decentralised wind turbines on a weak grid could reduce transmission losses and may reduce the need for grid reinforcement.

- Wind power is particularly useful for providing a local supply or in isolated communities.

Constraints

- Wind technology needs further development to reduce costs.

- The technology needs to be demonstrated in the UK to assess the environmental impact.

- The technology also needs the economies of scale and experience that can only be obtained from an expanding market.

- The variability of the wind speed results in a variable electricity supply which progressively reduces the value of the marginal unit of capacity as total installed capacity increases. Due to its intermittency it is considered that supplying more than about 10% of total electricity supply from wind energy would result in additional costs to the grid operating system.

- Lack of availability of good sites near centres of electricity demand.

Prospects and Categorisation

Onshore Wind		
Occurrence	New deployment in several scenarios in the short term	
Year Contribution (TWh/y) CO_2 savings (MtC/y)	2005 0.2 to 30 0.03 to 6	2025 0 to 30 0 to 6
Categorisation	Market enablement via NFFO and RDD&D	

Offshore Wind		
Occurrence	No deployment under the scenarios considered to 2025	
Year Contribution (TWh/y) CO_2 savings (MtC/y)	2005 0 0	2025 0 0
Categorisation	Watching Brief	

Table 3: Prospects and Categorisation

The Programme

Aims

1. To encourage the uptake of wind energy by:

   - assessing when the technology will become cost effective;

   - stimulating the development of the technology as appropriate;

   - establishing an initial market via the NFFO mechanism;

   - removing inappropriate legislative and administrative barriers;

   - ensuring the market is fully informed.

2. To encourage internationally competitive industries to develop and utilise capabilities for the domestic and export markets, taking account of what influences business competitiveness.

3. To quantify environmental improvements and disbenefits associated with wind energy.

4. To manage the programme effectively.

Justification

Within the UK, onshore wind energy has the potential to make a significant cost effective contribution to UK energy supplies, possibly as much as 10% of electricity generation, provided it is given credit for its environmental benefits.

With Government support an infant industry has been created in the UK which can be sustained by a programme of the size and scope envisaged and will provide a sound basis for the efficient and timely exploitation of market opportunities when they emerge.

Hydro Power

The Technology

Description and Present Status

Hydro power comes from the energy available from water flowing in a river or in a pipe from a reservoir. Evidence of the use of hydro power as a source of energy has been found in primitive devices from the first century BC. During the Industrial Revolution, small-scale hydro power was commonly used to drive mills and various types of machinery. The first large-scale hydro scheme was built in Scotland in 1896.

In modern times, hydro power is usually extracted using a turbine to generate electricity. The power available at a site is determined by the volume of water flowing and the hydraulic head or water pressure. A typical hydro scheme consists of the following items:

- a suitable rainfall catchment area
- a hydraulic head
- a water intake placed above a weir or behind a dam
- a method of getting the water from the head to a turbine
- a turbine, generator and associated buildings
- an outflow, where the exhaust water returns to the main flow.

Typically, larger schemes also include a reservoir providing seasonal storage to match the production of electricity to the demand for power from the area grid.

Hydroelectric technology can be regarded as being fully commercialised. Turbine plant, engineering services and turnkey systems are sold by UK and overseas organisations. Numerous schemes have been built in the UK, ranging from installations producing less than 1 kW to more than 100 MW. Hydroelectric schemes fall into two broad categories - large and small scale. Large-scale schemes are considered as those with an installed capacity in excess of 5 MW and were built by the electricity utility companies.

The Government's programme for small scale hydro-power has been focused on two main areas:

- stimulation of the wider take-up of existing commercially available small scale hydro in relevant market sectors, including water companies, industry and private individual operators;
- support of RD&D on novel technologies for extracting energy from low-head sources to establish whether or not they are technically or commercially viable.

A technical and economic assessment of the UK resource for small scale hydro has been completed and the results have been disseminated. These are now being used by developers to locate suitable sites.

A study of the non-technical barriers to the exploitation and commercial development of small scale hydro power has been carried out. The results have been published and made available to the relevant parties. Some of the obstacles identified in this study have now been removed, and those that remain have been highlighted. One problem identified by the study is the initial financial outlay necessary to determine whether or not a development is feasible. The Government has therefore been contributing 50% towards the cost of various feasibility studies for small scale hydro power sites. A lack of technical and procedural knowledge (planning requirements, electricity sales etc.) for small scale hydro power on the part of developers was also identified as a reason why schemes failed to be completed. Planning and technical guidance documents are being produced to meet this need.

The programme has also funded a number of research projects on novel concepts for the generation of electricity at sites with low hydraulic heads. This work is now largely complete. None of these innovations has yet proved to be attractive enough to take to the development stage.

In addition the DTI has an ongoing programme to assess the available resource and the likely cost of exploiting it.

DTI/DEn Programme Funding to 31/3/93	£9.4M
Number of R&D Projects to 31/12/93	24
Number of Commissioned NFFO Projects MW DNC to 31/12/93	27 19

Table 1: Estimated programme spend (money of the day) and number of projects.

Resource

Most of the large-scale hydro capacity in the UK is installed in Scotland (1.1 GW), with a smaller amount in Wales (140 MW). The small-scale sites comprise about 58 MW in Scotland and 20 MW in England & Wales. Hydro power accounts for about 2% of the total installed generating capacity in the UK. An additional category is low head hydro (less than 3m which is also categorised as small scale).

Within the UK there is little scope for further development of large scale hydro because of the cost and concerns about its environmental impact. However, the two scales of technology can extract energy from a rainfall catchment area in different ways. There is therefore a potential for the further development of small scale hydro for which approximately 80% of the unexploited resource is in Scotland.

	Accessible Resource (TWh/year) at less than 10p/kWh (1992), 8% Discount Rate.	Maximum Practicable Resource (TWh/year) in 2005 at less than 10p/kWh (1992), 8% Discount Rate.
Small scale hydro	3.9	3.9
Large scale hydro	6.9	3.9

Table 2: UK Hydro Resource - existing schemes provide the bulk of this resource (see ref 3).

Environmental Aspects

Hydro schemes are non-polluting, with no emissions of noxious gases, solid or liquid waste or heat. The systems are simple to decommission at the end of their lifetime. The plant and seasonal operating schedule can be designed in co-operation with the salmon and trout fishing industry to minimise any injury to fish and their smolts. Some types of turbine lead to increased oxygenation of the river water, to the benefit of the fish, and the installations collect and remove water-borne debris during operation. Small scale hydro schemes have been developed in sensitive landscape areas by minimising their visual impact. Large scale hydro schemes involving large dams and reservoirs have a significant environmental impact. The hydropower from a catchment area can also be harnessed by a series of small scale river schemes with less environmental impact but with a reduced storage and generation capacity.

Economic Prospects

The key aspects of the economics of a hydro power development are:

- a large initial capital outlay
- a long lifetime for the scheme
- high reliability and availability
- low running costs
- no annual fuel costs.

Costs are considerably reduced if existing engineering works can be used. A typical 1MW small scale hydro scheme could generate electricity for 3.5 p/kWh assuming an investment cost of £1000/kW, operation and maintenance costs of £30/kW/year, a lifetime of 30 years, a load factor of 60% and a 15% required rate of return.

Opportunities and Constraints

Opportunities

- Hydro technology is well proven, with standard equipment available and with minimum technical risk compared with novel technologies. There is usually no problem due to turbine noise as the plant can be enclosed in a building or underground.
- The construction times for small-scale systems has made them suitable as NFFO-funded projects.
- The stimulation of a competitive domestic market should further enhance the competitiveness of UK companies in export markets
- Small scale hydro schemes should have low environmental impact
- The locations of hydro are frequently in remote hilly areas where there may be benefits to the grid from local generation.

Constraints

- Hydro schemes require the major component of their costs i.e. construction in advance of generation, but there are relatively low operation and maintenance costs.
- Multiple ownership of the land, fishing and water abstraction rights, and other non-technical issues can increase the complexity of exploiting the hydro resource.
- Large scale schemes which involve a water storage reservoir have a significant environmental impact, and may not be acceptable, especially in areas of high landscape quality.
- Currently the rate of uptake of hydro potential is limited by the its cost and the absence of a guaranteed market for the electricity that could be generated.

Prospects and Categorisation

Small scale hydro		
Occurrence	New deployment in several scenarios	
Year Contribution (TWh/y) CO_2 savings (MtC/y)	2005 0.8 to 2.8 0.16 to 0.56	2025 0.8 to 2.8 0.16 to 0.56
Categorisation	Market enablement via NFFO and RDD&D	

Large scale hydro		
Occurrence	No new deployment under the scenarios considered to 2025	
Year Contribution (TWh/y) CO_2 savings (MtC/y)	2005 3.4 0.68	2025 3.4 0.68
Categorisation	Watching Brief	

Table 3: Prospects and Categorisation

The Programme

Aims

1. To encourage the uptake of hydro power by:

   - assessing when the technology will become cost effective;

   - stimulating the development of the technology as appropriate;

   - establishing an initial market via the NFFO mechanism;

   - removing inappropriate legislative and administrative barriers;

   - ensuring the market is fully informed.

2. To encourage internationally competitive industries to develop and utilise capabilities for the domestic and export markets, taking account of what influences business competitiveness.

3. To quantify environmental improvements and disbenefits associated with hydro power.

4. To manage the programme effectively.

Justification

Hydropower technology is well developed and there is at least 700 MW of unexploited hydroelectric power in the UK. It is a near market technology and the programme is intended to identify the resource, encourage the removal of remaining institutional constraints and demonstrate its environmental acceptability in order to assist the market should it wish to undertake hydropower projects.

Tidal Power

The Technology

Description and Present Status

A tidal barrage would be a major construction project built across an estuary, consisting of a series of gated sluices and low-head turbine generators. Several locations around the world have been studied as potential barrage sites, but relatively few tidal power plants have been constructed. The first and largest (240 MW) tidal plant was built in the 1960s at La Rance in France, and has now completed more than 25 years of successful commercial operation. There is no electricity generated from tidal energy in the UK.

The energy obtainable from a tidal scheme varies with location and time. The available energy is approximately proportional to the square of the tidal range, and the output changes not only as the tide ebbs and floods each day but can vary by a factor of four over a spring-neap cycle. However, this output is exactly predictable in advance. Extraction of energy from tides is considered to be practical only at those sites where the energy is concentrated in the form of large tides and in estuaries where the geography provides suitable sites for tidal plant construction. Such sites are not commonplace, but a considerable number have been identified in the UK. For example, the Severn Estuary where the maximum amplitude is about 11 metres.

Power generation during ebb tide conditions is the simplest mode of operation of a tidal system, although one can also increase the energy output by a small amount using the turbines in reverse to pump water into the basin at high tide.

The main objective of the Government's programme was to reduce uncertainty on costs, technical performance and effects on both the region and the environment to the point where it would be possible to make decisions on whether or not to plan for construction of specific barrage projects. The work being undertaken to meet this objective included studies on:

- the Severn and Mersey Barrages;
- small scale tidal energy;
- generic engineering issues;
- generic environmental issues.

Since the publication in June 1988 of Energy Paper 55, 'Renewable Energy in the UK - The Way Forward', substantial progress has been made.

A major (£4.2 M) development study for the Severn barrage was completed and reported in 1989.

Three stages of the Mersey Barrage feasibility study have also been successfully completed and reported bringing the total value of work to date to £7.2 M. A final report was produced in the spring of 1993.

Initial prefeasibilty studies for small scale schemes on the Loughor (5 MW), Conwy (33 MW) and Wyre (64 MW) have also been completed, and a similar study for a barrage on the Duddon (100 MW) is approaching completion.

Government has also continued to support a range of generic studies. These have underpinned the commercial site-specific barrage studies, by exploring possible methods of lowering costs and reducing technical and environmental uncertainties, and have mostly been fully funded by Government.

The R&D programme has now identified the likely costs of exploiting the resource. At the best sites electricity could be produced at about 7p/kWh at an 8% discount rate, and around 14p/kWh at a 15% discount rate (1991 prices). Since the technology associated with tidal energy barrages is reasonably well proven, there is little potential for substantially reducing these costs.

A recent assessment of tidal stream energy suggests energy could be extracted from coastal currents but with great difficulty. For simplicity of operation in a hostile marine environment and with low energy density, large numbers of fixed speed, fixed orientation, bottom mounted axial flow turbines are likely to be the most practical option. A resource estimate of 58 TWh/y based on arrays of such devices suggests that electricity might be produced, but at costs between 10p/kWh and 204 p/kWh at an 8% discount rate and 16p/kWh to 336p/kWh at a 15% discount rate (1992 prices).

Given the costs of exploiting tidal energy in the UK, in July 1993 the Department announced its intention to complete the Tidal programme.

DTI/DEn Programme Funding to 31/3/93	£12.1M
Number of R&D Projects to 31/12/93	150
Number of Commissioned NFFO Projects to 31/12/93	0

Table 1: Estimated programme spend (money of the day) and number of projects.

Resource

The UK has probably the most favourable conditions in Europe for generating electricity from the tides. This is the result of an unusually high tidal range along the west coast of England and Wales, where there are many estuaries and inlets which could be exploited. Tidal ranges in Scotland and Northern Ireland are too low for viable tidal energy schemes which would be considerably less economic than the most promising sites such as the Severn.

The Accessible Resource for tidal energy, regardless of cost, can be estimated by including every reasonably practicable estuary in the UK with a mean spring tidal range exceeding say 3.5 metres. This yields about 50 TWh/y, which represents about 20% of the present electricity consumption in England and Wales. About 90% of this would be at eight larger sites (Severn, Dee, Mersey, Morecambe Bay, Solway Firth, Humber, Wash and Thames), while the remaining potential is distributed over about thirty smaller sites. The Severn Barrage, if built, would be one of the largest civil engineering projects in the world, providing about 17 TWh/y of electricity, which would make up the majority of the Accessible Resource of 18.6 TWh/y at less than 10p/kWh (1992 prices) using an 8% Discount Rate.

Accessible Resource (TWh/year) at less than 10p/kWh (1992), 8% Discount Rate.	Maximum Practicable Resource (TWh/year) in 2005 at less than 10p/kWh (1992), 8% Discount Rate.
19	1.6

Table 2: UK tidal power resource (see ref 3).

Environmental Aspects

An important factor which will influence whether a tidal scheme can proceed is its likely effect on the environment and on the region. A thorough study of aspects such as water quality, sediment regime, bird and fish populations and the likely effect of the barrage on these must be undertaken on a site-by-site basis. Close attention must be paid to maintaining effective communications with the local population and providing sound information about the details of the scheme and its expected results.

There are several advantages arising from the construction of a tidal barrage in addition to providing a clean, non-polluting source of renewable energy. Tidal barrage schemes can assist with the local infrastructure of the region, create regional development opportunities and provide protection against coastal flooding within the basin during storm surge (exceptionally high) tides.

Economic Prospects

Tidal power incurs high capital costs per kilowatt of installed capacity when compared with other electricity generation technologies. Construction times can be several years for the largest projects. Operation is also intermittent, with a relatively low load factor, typically around 23%. These characteristics tend to lead to high generation costs. Total annual charges, including operation and maintenance costs, are low, typically 0.5% of the initial capital cost. With appropriate maintenance, the lifetime of a scheme can be very long (possibly greater than 120 years for the main barrage structure and 40 years for the plant and machinery).

The economics of tidal energy depend primarily on site-specific factors. UK studies have identified promising schemes with capacities ranging from 30 MW to 8 GW which have broadly similar energy costs. There is no significant economy of scale. The best estuaries, including the Severn, would produce electricity at about 7 p/kWh, if construction could be undertaken with capital costs discounted at 8% per annum. With construction costs discounted at 15% per annum, the cost of electricity would double to about 14 p/kWh, at 1991 prices. Estimates of additional non-energy benefits are of secondary importance and believed to be minor.

Opportunities and Constraints

Opportunities

- Tidal energy represents an opportunity for renewable energy generation and reduction in carbon dioxide emissions on a large scale.

- There are non-energy benefits which can result from building a barrage (such as the possibility of providing road crossing points, increased local employment).

- There could be export opportunities for UK consultants in tidal project design and assessment.

Constraints

- There is no deployment of tidal energy under any of the future scenarios which have been envisaged. This is because of the economics of electricity generation from tidal barrages.

- Tidal barrage schemes could have significant environmental effects because they would inevitably cause changes to estuarine ecosystems. A site-specific environmental impact assessment would be necessary at each proposed location.

- The relatively high capital costs and long construction period of the larger schemes makes the unit cost of electricity sensitive to discount rate. Non-energy benefits are of secondary importance.

Prospects and Categorisation

Tidal Power		
Occurrence	No deployment under the scenarios considered to 2025	
Year Contribution (TWh/y) CO_2 savings (MtC/y)	2005 0 0	2025 0 0
Categorisation	Watching Brief	

Table 3: Prospects and Categorisation

The Programme

Aims

1. Complete the existing programme.

2. Maintain a watching brief on the technology.

Justification

The generating costs of electricity from tidal energy schemes are now well understood. Given these costs, tidal energy barrages will not be commercially developed under any of the future scenarios considered. Much work has already been done, both generically and on site-specific schemes to reduce technical, economic and environmental uncertainties. It is currently believed however, that there is only limited scope for cost reduction.

It is appropriate now to complete the programme in such a way as to maximise the value of projects currently under way and to enable work to be picked up at a later stage should circumstances change significantly beyond the scenarios studied.

Wave Energy

The Technology

Description and Present Status

Ocean waves are caused by the transport of energy from winds as they blow across the surface of the sea. The amount of energy transferred depends upon the speed of the wind and the distance over which it acts. As deep ocean waves suffer little energy loss, they can travel long distances if there is no intervening land mass. Therefore the western coastline of Europe has one of the largest wave energy resources in the world, being able to receive waves generated by storms throughout the Atlantic.

Wave energy is still in the R,D&D phase. Currently there are two types of device known to be operating in Europe, an oscillating water column on Islay in Scotland and a tapered channel (Tapchan) in Norway. The latter fills a reservoir sited about 5 metres above mean sea level through a funnelling channel and generates power by returning the water through a Kaplan turbogenerator into the sea. This concept is limited to areas where there is a small tidal rise and fall and suitable shoreline topography.

World-wide, installed devices are limited to experimental plants of less than 100 kW, including oscillating water column devices incorporated in sea defence breakwaters.

The UK R&D Programme from 1974 to 1983 spent over £17M examining more than 300 design concepts for converting wave energy into useful power. Eight of these concepts were developed and costed as full scale 2 GW electrical generating systems. The principal energy conversion concepts were tested in specially constructed experimental facilities. Large scale models were also tested at sea and in large lochs. Since 1985, Government has funded work on a 75 kW oscillating water column device incorporating a Wells air turbine developed by the Queens University of Belfast (QUB). This device comprises a concrete chamber astride a natural rock gully on the seashore. Waves entering the gully cause air in the chamber to vent and return by way of a turbine. The uni-directional Wells turbine extracts energy from the air as it flows in either direction, the turbine continuing to rotate in the same direction. This device is now connected to the grid on Islay in the Inner Hebrides.

In 1992 the results of a review of the technical feasibility and commercial viability of wave energy, carried out with the assistance of both the device teams and independent consultants, were published - Review of Wave Energy, Ref 4. Five major and three more recent less well developed designs as well as the wave energy resource were studied. It was concluded that the main devices assessed were unlikely to generate electricity competitively in the short to medium term. There might be some scope for a further reduction in generating costs beyond those calculated but, for most of the devices, this would require significant changes to the designs considered in the review.

DTI/DEn Programme Funding to 31/3/93	£19.1M
Number of R&D Projects to 31/12/93	153
Number of Commissioned NFFO Projects to 31/12/93	0

Table1: Estimated programme spend (money of the day) and number of projects.

B14

Resource

The principal offshore wave energy levels for the British Isles are of the order of 60 to 80 kW/m of wavefront. The energy levels are high because of the long distance across the Atlantic where the winds interact with the ocean surface. The shadowing effects of Ireland on the coast of England and Wales are, however, significant.

The UK wave energy resource can be subdivided as:

- shoreline - where the device is constructed on the seashore

- nearshore - where the device is floating or bottom mounted in 10 to 25m water depth

- offshore - where the device is moored in greater than 40m water depth.

The wave energy levels decrease from the values quoted above as the shore is approached, due to sea bed friction and other effects.

The major constraints on the full exploitation of offshore and nearshore devices would be the effects on shipping lanes and access to fisheries, combined with the need for sea to shore power cables. The associated shoreline grid terminations may raise environmental objections. The restraints on shoreline devices may arise from objections to the impact of such structures on natural coastlines.

	Accessible Resource (TWh/year) at less than 10p/kWh (1992), 8% Discount Rate.	Maximum Practicable Resource (TWh/year) in 2005 at less than 10p/kWh (1992), 8% Discount Rate.
Shoreline	0.4	0.25
Offshore	0.03	0.03

Table 2: UK Wave Energy Resource (see ref 3).

Environmental Aspects

Wave energy is a non-polluting energy source, producing no noxious emissions. Whilst all three categories would require land based sites to make the necessary grid connections, the impact of such facilities would be low but will need evaluation before deployment. The associated power transmission lines would be no more intrusive than for any other form of generation.

Shoreline plants must be sited in exposed situations, but need not be visually intrusive and would not be expected to exceed a land utilisation of 200 sq.m per MW. However such sites must be considered as environmentally and visually valuable and would require grid connection and access roads.

Nearshore devices have little land use, but will have similar environmental impact to shoreline plant.

Offshore devices will have minimal visual impact but may have other varied effects, either beneficial or detrimental, depending on the choice of site. These might include, new sheltered areas, havens for fish or birdlife, effects on coastal wave energy and hence on resident organisms and possible pollution from lubricants or anti-fouling materials. Navigational hazards could possibly arise from the failure of moorings etc. Techniques for the recovery of such hazardous equipment have been demonstrated by the oil industry.

Noise from air turbines may be an issue for nearshore and shore based plant.

Economic Prospects

The costs of two representative wave energy devices, studied in the 1992 Wave Energy Review, are presented here. These are a 1 MW shoreline gully oscillating water column (OWC) and a 12.5 MW scheme of 5 circular clams for offshore deployment.

The estimated cost of a shoreline OWC device in the Review was £1,530/kW with annual operating costs of £44/kW. The estimated cost a circular clam scheme was £2,600 to £3,000 per kW with annual operating costs of £66/kW.

The data are in 1992 values with an inflation factor of 1.09 on the 1990 data published in the Review. In order to address the relatively undeveloped nature of wave energy, the Wave Energy Review gave very considerable allowance for the potential benefits of R&D on the practicality, performance and costs of the wave energy devices considered.

On the basis of the findings of the Wave Energy Review, the existing large scale offshore device designs are unlikely to become competitive by 2025 under any of the future energy scenarios studied. The shoreline wave energy resource becomes economic under a few of these scenarios if all R&D is successful. The estimated contribution is limited and amounts at most to 0.16 TWh/year.

Opportunities and Constraints

Opportunities

- Wave energy could provide a non-polluting power source providing the costs can be made competitive.

Constraints

- Uncertainties concerning the technology, performance, reliability and consequently the costs deter commercial interest at present.

- The UK shoreline resource is limited by geological and other constraints

- The technical performance and reliability of many components over a long term in a marine environment should be demonstrated.

- The applicability of conventional offshore mooring technology has to be established for floating devices and seabed emplacement techniques for fixed devices must be developed and demonstrated. These techniques have been developed for shipping, barges and offshore oil platforms.

Prospects and Categorisation

Shoreline Wave Energy		
Occurrence	New deployment in a few scenarios within the long term	
Year Contribution (TWh/y) CO_2 savings (MtC/y)	2005 0 to 0.16 0 to 0.032	2025 0 to 0.16 0 to 0.032
Categorisation	Watching Brief	

Offshore Wave Energy		
Occurrence	No deployment under the scenarios considered to 2025	
Year Contribution (TWh/y) CO_2 savings (MtC/y)	2005 0 0	2025 0 0
Categorisation	Watching Brief	

Table 3: Prospects and Categorisation

The Programme

Aims

1. Complete the existing programme.

2. Maintain a watching brief on the technology.

Justification

Whilst the UK has a large theoretical wave energy resource, current offshore device designs are not economic. There might be scope for some reduction in cost but substantial modification in existing designs, or radically new device concepts, would be necessary. Even if successfully developed, and assuming successful completion of all outstanding R&D, any contribution would be so limited that further R&D is inappropriate.

Photovoltaics

The Technology

Description and Present Status

Photovoltaic (PV) materials generate direct current electrical power when exposed to light. Power generation systems using these materials have the advantage of no moving parts and can be formed from thin layers (1 - 250 microns) deposited on readily available substrates such as glass. To date, the photovoltaic effect has been exploited where the low power requirements, good solar resource and simplicity of operation outweigh the high cost of PV systems. Current applications include consumer goods, such as calculators and watches, powering of systems such as remote telecommunications facilities and, on a larger scale, power systems for lighting and water pumping in developing countries and in remote areas with no grid supply.

PV is still a relatively young technology. Much research and development will continue to be necessary if world-wide system costs (modules and associated components) are to be reduced to acceptable levels and significant new markets are to be established. Current trends in module cost reductions suggest that this is a realistic goal.

PV could contribute to electricity supply in two ways - through the use of central PV generating plant (PV power stations) or through building-integrated systems where PV units would be located in the facades of domestic and commercial buildings. These could supply power for use inside the buildings with any excess available for export via the grid. At present, the economic prospects for central PV generating plant look poor for the UK. However, building-integrated systems might become an economic proposition in the future, even under UK climatic conditions, if the PV system can be incorporated into the building structure, displacing conventional building materials and components.

Photovoltaic systems were not included in Energy Paper 55. A recent review of the technology showed that significant developments have occurred in terms of both system performance and unit manufacturing costs. As a result of this a programme on PV systems has been instigated within the DTI programme. The objectives of this initial programme on PV can be summarised as follows:

- to assess the potential for distributed PV electricity generation in the UK

- to identify the barriers to the installation and use of PV systems.

Without Government involvement, the development of PV systems for building integration is likely to be very slow. The main reasons for this are the conservatism of the construction industry with regard to new technologies and the perceived risks of applying new components and techniques. New products will need to be developed for this market which will require investment by the PV and building industries. In addition, certification and testing procedures will need to be established. There is a need therefore, to determine what the technical and market barriers facing this application are likely to be. If the energy, environmental and wealth creation prospects for PV technology can be shown to be favourable then the relevant industries might be encouraged to make the required investments.

DTI/DEn Programme Funding to 31/3/93	£0.3M
Number of R&D Projects to 31/12/93	7
Number of Commissioned NFFO Projects to 31/12/93	0

Table 1: Estimated programme spend (money of the day) and number of projects.

Resource

The horizontal daily insolation (averaged over the year) in the UK ranges from 2.2 kWh/sq.m/day in northern Scotland to 3.0 kWh/sq.m/day in southern England. However, the efficiency of the photovoltaic conversion process for current commercial modules ranges between 5% and 16%, so that only a fraction of the radiation received can be utilised (although the best laboratory PV cells have measured efficiencies greater than 30% under concentrated sunlight). The usable resource is also constrained by the land area that can be made available for PV systems. For central PV generating plant, the maximum area available would be determined by planning considerations after allowing for farmland, roads, lakes, forests, national parks etc. For building-integrated systems the Accessible Resource will depend on the area of south-facing roofing and walls which could be fitted with PV systems.

At present, the economics of central PV generating plant look unpromising in the UK, even well into the future, and therefore no detailed assessments have been made of their land use requirements. For building-integrated systems, the realistic resource which might be exploited is likely to be that fraction of newbuild roof and wall areas where PV components can be built in. This will depend on PV components and support structures displacing conventional building materials and components at a cost acceptable to the building market. No significant new contribution is expected by 2005 using a 15% discount rate.

The exploitable PV resource depends heavily on the conversion efficiency of the system and its overall cost. Most current off-the-shelf modules have a cost of about £2 per peak Watt output. It is thought that this could be halved over the next 10 years. New materials which promise higher efficiencies or lower costs than the crystalline and amorphous silicon cells currently used are under development in several countries.

Accessible Resource (TWh/year) at less than 10p/kWh (1992), 8% Discount Rate.	Maximum Practicable Resource (TWh/year) in 2005 at less than 10p/kWh (1992), 8% Discount Rate.
84	0.004 to 0.08

Table 2: UK Photovoltaics Resource (see ref 3).

Environmental Aspects

Widespread use of PV offers the prospect of reduced emissions associated with fossil-fuelled power stations. Furthermore, building-integrated PV systems create no noise or pollution and need not be visually intrusive. However the manufacture of PV components involves the use of some hazardous materials. In general, these can be coped with adequately using standard industrial treatment techniques such as multi-stage scrubbers. None of the adverse environmental effects associated with the manufacture, use or disposal of PV materials appears to present major technical difficulties or raise new issues of regulatory control. The large-scale deployment of PV technology, particularly in the form of central generating plant, could have land use and therefore planning consent implications.

Economic Prospects

Even with optimistic assumptions about system efficiency and resource availability, the cost of electricity from central PV generating stations in the UK is likely to be above 10p/kWh (1991 prices) in the year 2025.

Building-integrated systems offer better prospects. They avoid transmission losses and achieve reduced overall costs by removing the need for support structures and replacing facade and roofing materials by PV systems. Present estimates suggest that fully developed building-integrated systems based on improved technology might achieve electricity generation costs of less than 10p/kWh (1991 prices).

For distributed PV systems, it is possible that the value of the power could be greater than implied by the generation costs, depending on how much PV power could be utilised directly by the building operator and the tariffs of the displaced mains electricity.

Opportunities and Constraints

Opportunities

- PV technology is a modular, silent and pollution free potential energy generation source which is attracting widespread market pull for more applications of PV systems.

- Many buildings are increasingly becoming major users of daytime electrical energy for lighting and air-conditioning. Trends towards flexible tariffs and intelligent metering of supplies will increase desire for the local generation and control that PV systems can provide.

- New PV materials currently under investigation together with an increased world market promise major reductions in module cost from the current £2/Wp to £1/Wp and perhaps to £0.3/Wp in the longer term.

- PV technology is attracting substantial funding world-wide from both public and private sources. The DTI programme will provide a base for the UK to attract some of this funding, from the CEC in particular, and significantly augment the DTI contribution.

Constraints

- PV systems for electricity supply are currently very expensive, and large expenditure over the next two decades will be necessary on R&D, demonstration and manufacturing facilities if target costs and performance are to be achieved.

- PV will only be widely adopted in the UK if and when small-scale distributed systems become an acceptable means of generating electricity for domestic and commercial use.

- Some PV system materials may pose environmental hazards.

- There are no market incentives to stimulate the deployment of this technology.

Prospects and Categorisation

Photovoltaics		
Occurrence	New deployment in a few scenarios in the long term	
Year Contribution (TWh/y) CO_2 savings (MtC/y)	2005 0 to 0.03 0 to 0.006	2025 0 to 7.2 0 to 1.4
Categorisation	Assessment and RDD&D	

Table 3: Prospects and Categorisation

The Programme

Aims

1. To assess photovoltaics maintaining the option of developing and deploying at a later stage.

2. To encourage internationally competitive industries to develop and utilise capabilities for the domestic and export markets, taking account of what influences business competitiveness.

3. To quantify environmental improvements and disbenefits associated with photovoltaics.

4. To manage the programme effectively.

Justification

Photovoltaic technology is one of the most attractive of the renewable energy sources in terms of a large Accessible Resource, potential for cost reduction, simplicity of operation, flexibility of deployment and small environmental impact. There is a major world-wide effort at the R&D and manufacturing process level to develop systems of suitable performance and cost for widespread deployment. A DTI programme would enable the UK to benefit from this investment at only a small cost.

Photoconversion

The Technology

Description and Present Status

Photoconversion is a term for various processes which convert sunlight directly into either electrical power, heat or a chemical fuel. These include photobiological, photochemical and photoelectrochemical processes. The main attractions as an energy source are:

- they rely on the Sun's energy, which is freely available, to drive the reactions;

- they often use water as a substrate for many of the required reactions;

- they have low ecological impact;

- they do not add to atmospheric CO_2 levels.

Photoconversion technology is still at the laboratory research stage. Several photoconversion processes are being investigated world-wide, but at present only two are thought to be worthy of attention as possible future energy options for the UK - electrochemical photovoltaic cells (ECPV) and photobiological systems to produce hydrogen.

Electrochemical photovoltaic cells are functional equivalents of conventional solid-state semiconductor photovoltaic (PV) cells, producing direct electrical current (DC) on illumination. They are far less developed than solid-state PV cells but their cell efficiencies are comparable. Indeed, ECPV cells might eventually have some advantage through lower production costs.

The main difference between the PV and ECPV systems is that the light-reactive centre of photovoltaic cells is embedded within the semiconductor, whereas it is on the electrode surface in ECPV cells. This means that there is a tolerance to a rough electrode surface which renders ECPV design potentially inexpensive in comparison to a photovoltaic device where effectiveness depends upon a uniform and accurate junction inside the semiconductor crystal matrix.

Many laboratory-scale systems capable of continuous hydrogen production have now been developed in various parts of the world. The efficiency of the conversion of solar energy into hydrogen by anaerobic photosynthetic bacteria has been estimated at between 5 and 6%. The National Renewable Energy Laboratory in the USA believes that, with optimisation, efficiencies of 10-15% could be achieved.

The UK Government funds fundamental research through the Research Councils mainly based at universities. To date photoconversion has not been included in the DTI's New and Renewable Energy Programme, although a review has been undertaken to assess the potential of photoconversion as an energy production process in the UK. The review included both scientific assessment and engineering implications. Most research funds, both in the UK and internationally, are concentrated at the fundamental investigative level. There is currently no mechanism for taking the results of these basic studies forward to a pilot scale 'demonstration of the theory' phase.

DTI/DEn Programme Funding to 31/3/93	£0 M
Number of R&D Projects to 31/12/93	0
Number of Commissioned NFFO Projects to 31/12/93	0

Table 1: Estimated programme spend (money of the day) and number of projects.

Resource

The resource for photoconversion systems will be determined by the solar energy available, the system efficiencies and the way the systems are applied. At present there is insufficient information for any detailed resource assessment to be made. However there do not appear to be any over-riding problems associated with the characteristics of the UK solar resource which would, in principle, preclude the operation of efficient photoconversion systems in this country.

Electrochemical photovoltaic cells would probably be deployed in a similar manner to building-integrated photovoltaic systems. Thus the resource will be determined by the amount of roof, wall and other areas that could be made available for housing the ECPV cells.

Photobiological hydrogen systems would be best utilised in parallel with other waste treatment processes. The application of this technology will therefore be limited by the economics of adding it to existing installations with suitable growth mediums.

Accessible Resource (TWh/year) at less than 10p/kWh (1992), 8% Discount Rate.	Maximum Practicable Resource (TWh/year) in 2005 at less than 10p/kWh (1992), 8% Discount Rate.
84	unknown

Table 2: UK Photoconversion Resource (see ref 3).

Environmental Aspects

No detailed environmental analysis has yet been attempted for photoconversion systems. On the benefit side, both photobiological and electrochemical systems have the potential for providing electricity without the CO_2 and harmful emissions, such as NO_x and SO_x, produced by fossil-fuelled plant. On the debit side, there may be some environmental burdens associated with wastes from the production and use of photochemical materials, but these are not unique to photoconversion processes and should not hinder exploitation. It is not thought that there will be any planning considerations specific to photoconversion technologies.

Economic Prospects

Since this technology is still at the research end of the R&D stage, there are as yet no prototype systems upon which initial economic analyses could be based. The speculative cost, for an installation producing 10 MWe, is between £2.5m and £5m per MWe. This is based on current understanding of how the technology is likely to develop and upon costs for analogous solar power systems. For systems with a 25 year lifetime this might lead to costs of about 20p/kWh at a discount rate of 8%.

Opportunities and Constraints

Opportunities

- There is an ongoing development of fundamental photoconversion science carried out by the SERC in the UK and various bodies overseas. The programme would benefit from this knowledge base, and choosing key areas, build upon it to develop proto-type schemes which could attract industrial involvement.

- Some processes can benefit from developments in other energy technologies, e.g. solid state photovoltaics developments will support photo electrochemical systems and biogas work will benefit photo biological systems.

- Like other solar energy technologies, photoconversion processes offer promise of an alternative method of producing electricity free from the harmful emissions associated with fossil-fuelled plant.

- ECPV cells could, in principle, contribute to pollution-free electricity supply through distributed systems located on buildings. In the long term, they could prove more economic than conventional PV cells.

- Photobiological hydrogen systems offer the prospect of additional methods of waste processing with energy production and environmental benefits.

Constraints

- The lack of knowledge of how systems scale up from laboratory processes.

- The lack of quantified data to allow resource and cost assessment.

- The uncertainty over the environmental impact of some processes.

Prospects and Categorisation

Photoconversion		
Occurrence	Insufficient data to allow an assessment	
Year Contribution (TWh/y) CO_2 savings (MtC/y)	2005 unknown unknown	2025 unknown unknown
Categorisation	Assessment and RDD&D	

Table 3: Prospects and Categorisation

The Programme

Aims

1. To assess photoconversion technologies maintaining the option of developing and deploying at a later stage.

2. To encourage internationally competitive industries to develop and utilise capabilities for the domestic and export markets, taking account of what influences business competitiveness.

3. To quantify environmental improvements and disbenefits associated with photoconversion technologies.

4. To manage the programme effectively.

Justification

Photoconversion is a generic term covering a number of processes for converting sunlight energy into electricity, heat or chemical fuels. The majority of these technologies can be grouped under the headings of photo chemistry, photo electrochemistry and photo biology. Initial assessments of a broad range of these technologies suggests that several offer prospects of long term application in the UK. A more detailed and focused programme of assessment is required to enable their potential to be properly determined.

Active Solar

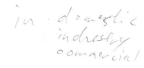

in domestic
industry
commercial

The Technology

Description and Present Status

Active solar thermal systems consist of solar collectors, which transform solar radiation into heat, connected to a heat distribution system. Due to the nature of the UK climate, such heating systems are best suited to applications at temperatures below 100°C. High temperature applications, such as electricity generation, are not practical in the UK. There is a developed technology and an existing, small market for systems to supplement the heating energy demands of domestic buildings. This market is served by a small number of manufacturers and installers, but many of the installers see solar heating as a secondary activity associated with another business, such as conventional heating system installation.

A solar domestic hot water (DHW) system is usually supported by a conventional water heating system using fossil fuels or electricity to ensure that the water temperature exceeds 60°C throughout the year. In a common configuration the solar heat is used to pre-heat the water in a conventional DHW system using a closed circuit calorifier. The working temperature is then attained using another heat source.

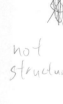

hot
structure

Solar collectors are commonly either black-surfaced flat plate collectors, usually mounted behind a clear glass sheet to reduce heat losses, or an assembly of evacuated tube collectors. High technology glazings have been developed to further reduce heat losses, and these may be applied in the future. The fluid in the flat plate collector is typically water containing a non-toxic anti-freeze additive but the primary fluids in evacuated tube collectors are more complex. Air-based collectors are a special case of the flat plate collector and use a mineral-based heat store, such as a pebble bed, to feed pre-heated air to air-based space heating systems.

A typical UK solar domestic hot water installation would use 3-5 sq.m of collector. If correctly designed and installed, it would be expected to supply 250 to 550 kWh/sq.m/y, depending on the hot water load of the household and the efficiency of the system. The projected UK manufacture of collectors for 1991 was 36,000 sq.m of which 24,000 sq.m was for export. The 1991 domestic sales therefore represent approximately 4.5 GWh/y of installed capacity. At least one UK company is a world leader in the development and manufacture of active solar systems in what can be considered a high technology industry.

The Government's Active Solar R&D Programme, carried out by the former Department of Energy between 1977 and 1984, coincided with a period of expanding world demand for active solar products. The Programme was designed to define the potential contribution of the technology to UK energy supplies and to stimulate the development of cost-effective systems.

Although several of the 70 or so projects examined other applications, the programme concentrated on the application of active solar heating to domestic water and space heating. Particular attention was paid to component and system development, laboratory testing, field trials and modelling studies.

The programme proved that active solar heating is technologically viable in the UK, particularly in low temperature applications.

Since 1984 the Government maintained a watching brief in the area until a review was completed in 1992. This review predicted a potential resource of around 12 TWh/y for solar domestic hot water heating alone. However, to encourage a mass market to develop it concludes that it will be important to raise public confidence in the systems by providing appropriate information and promotional material.

DTI/DEn Programme Funding to 31/3/93	£4.1M
Number of R&D Projects to 31/12/93	73
Number of Commissioned NFFO Projects to 31/12/93	0

Table 1: Estimated programme spend (money of the day) and number of projects.

Resource

The total resource for active solar depends upon the total incident solar radiation. The intensity of this insolation varies from 900 kWh/sq.m/y in northern Scotland to 1250 kWh/sq.m/y in south western England, assuming that the collector is tilted southwards at an angle of 30°.

The Accessible Resource for DHW systems will depend upon the number of suitably equipped and sited buildings, the collector type and the required heat load. Not all buildings can accommodate a correctly orientated collector.

	Accessible Resource (TWh/year of heat) at less than 10p/kWh (1992), 8% Discount Rate*.	Maximum Practicable Resource (TWh/year of heat) in 2005 at less than 10p/kWh (1992), 8% Discount Rate.
Heat for Domestic Hot Water	12	1.8
Heat for Swimming Pools	0.8	0.12
District Heating Schemes	18	5.9
Industrial and Commercial applications	Unknown	Unknown

Table 2: UK active solar resource (see ref 3).
* based on estimated costs for fully developed systems.

By comparison the currently installed capacity for active solar DHW systems is estimated as 36 GWh/y and for swimming pool heating as 28 GWh/y.

Cons

Environmental Aspects

Active solar heating is a truly renewable and emission free energy source. Its environmental impact is limited to the various manufacturing processes involved, including the normal production of iron, steel, glass etc.

The need to store hot water at temperatures in excess of 60° C, in order to discourage the growth of Legionella or other hazardous organisms, is important, but is equally applicable to conventional hot water systems based on fossil fuel or electricity heating. Only during hot sunny weather might a household expect its hot water requirements to achieve and retain this temperature on solar heating alone.

Economic Prospects

There is a large variation in the current prices charged for solar DHW systems installed in the UK. The consumer price for a typical system, including 17_% VAT, may vary from £1400 to £4700, with running costs of about £6 per year. This range includes both flat plate and evacuated tube systems of varying sizes, but most systems sold are in the middle to the upper regions of this range.

The energy costs for a 25 year life and an 8% Discount Rate are thought to typically range from 15 to 32p/kWh. However, with a 0% discount rate this would fall to 7 to 14p/kWh. It is unclear how best to assess the economics applicable to a domestic situation. In the existing UK market purchasers, do not buy for economic reasons alone. However for the market to expand substantially, this technology will have to become more economically attractive to the consumer.

Opportunities and Constraints

Opportunities

- Increased awareness by information dissemination could reduce marketing costs and help bring down system prices.

- There is the opportunity for further technical development of collectors and ancillary equipment in order to reduce costs, simplify installation and improve performance.

- There is an increased awareness in the consumer market of environmental issues. The programme will seek to exploit this by providing credible and independent information on active solar systems.

- The programme will make use of the ongoing national programmes in other countries and the IEA on Solar Aided District Heating Schemes (SADHS) development.

Constraints

- Reduction of overall costs is necessary if penetration of the mass market is to be achieved.

- The lack of public confidence and absence of informative material significantly increases the marketing costs.

- Concern over issues of planning, insurance and health risks.

- Lack of agreed international standards and a certification agreement.

- Lack of robust technical information.

Prospects and Categorisation

Active Solar		
Occurrence	New deployment in a few scenarios in the long term	
Year Contribution (TWh/y) CO_2 savings (MtC/y)	2005 0.01 (heat) 0.001	2025 0 to 2.5 (heat) 0 to 0.26
Categorisation	Assessment and RDD&D	

Table 3: Prospects and Categorisation

The Programme

Aims

1. To assess active solar technologies maintaining the option of developing them at a later stage.

2. To encourage internationally competitive industries to develop and utilise capabilities for the domestic and export markets, taking account of what influences business competitiveness.

3. To quantify environmental improvements and disbenefits associated with active solar technologies.

4. To manage the programme effectively.

Justification

There is an existing small market in the UK for Active Solar water heating systems and a much larger overseas market. The industry operates in the very difficult consumer products area where confidence in the technology and profile can be more important than strict cost effectiveness. The programme will increase the confidence in the technology at very little direct cost and sustain the current market and provide encouragement to industry to expand the home and overseas markets.

Solar Aided District Heating Systems have the potential to enable large scale exploitation of solar energy. At the present time there are two main barriers, the system costs are too high and (in the UK), district heating systems are not currently favoured. There are currently several large programmes in SADHS overseas and it may be appropriate to review the prospects for the technology in the UK when the results of these activities are known.

Passive Solar Design

The Technology

Description and Present Status

Passive Solar Design (PSD) aims to maximise free solar gains to buildings so as to reduce their energy requirements for heating or cooling and lighting. It is most effective when used together with energy efficiency measures as an integral part of energy-conscious design of new buildings. Some PSD features, such as conservatories and roof space collectors, can be retrofitted.

The concept of PSD is not new. However, its potential energy benefits, as distinct from its use for aesthetic or health reasons, have only relatively recently become a focus of attention.

To maximise these benefits in terms of the heating requirements of a building, PSD seeks to orientate and arrange glazed surfaces so as to make full use of shortwave solar radiation for heating interior spaces and to avoid heat loss resulting from siting windows on shaded walls. For the cooling of buildings it uses solar heated air to assist natural convection, thus providing natural ventilation and cooling. For lighting, it uses glazing to reduce the need for artificial lighting whilst still maintaining a comfortable environment. More complex approaches such as mass walls, atria or conservatories are basically extensions of these simple design principles.

Effective use of PSD depends on sympathetic interior design and on grouping buildings to minimise shading and gain protection from prevailing winds.

The passive solar programme has been assessing various passive solar design measures to identify develop and commercialise those which could be cost effective. The objectives for the programme, embodied in Energy Paper 55, were:

- to evaluate those aspects of the technology that could make significant and economic changes to the energy used in buildings;

- to develop and test design guidance so as to reduce the risks of applying the technologies;

- to produce information on the performance of passive solar design in practice;

- to put together messages about passive solar that will influence building designers, developers and users;

- to develop technology transfer tools to ensure that these messages are understood and taken up effectively.

Most of the objectives for these activities have now been achieved.

A considerable number of passive solar design studies have been undertaken and reported for both domestic and non-domestic buildings. Many of these have involved commercial collaboration with design teams, clients and builders. The results from the programme are being fed into the Energy Efficiency Office's Best Practice programme for energy efficiency measures in buildings

An energy design advice scheme is in place providing access to regionally based expertise on low energy design including PSD. This scheme should achieve annual energy savings of £50m by the year 2000.

DTI/DEn Programme Funding to 31/3/93	£14.1M
Number of R&D Projects to 31/12/93	125
Number of Commissioned NFFO Projects to 31/12/93	0

Table 1: Estimated programme spend (money of the day) and number of projects.

Resource

Although sunlight is a very variable resource in the UK, passive solar design in its simplest forms is used within virtually every building. Additionally, it can be easy to design buildings with the concept of PSD specifically in mind, and there are many examples of its incorporation in individual housing projects and a smaller number of non-domestic buildings. There is still little experience, for both technical and market-related reasons, of successful PSD applications in mass market housing schemes. However, there will always be a need for new and refurbished buildings which will continue to provide opportunities for low energy design.

It is difficult to quantify the size of the "installed capacity" presently resulting from PSD in the UK since there is much unplanned use of solar gain. In the UK, estimates suggest that such unplanned use contributes up to 145 TWh/y to energy requirements in buildings. There are presently some 500 houses and 100 non-domestic buildings in the UK specifically designed to maximise passive solar gain, yielding an estimated additional energy saving of 1.5 GWh/y.

Accessible Resource (TWh/year) at less than 10p/kWh (1992), 8% Discount Rate.	Maximum Practicable Resource (TWh/year) in 2005 at less than 10p/kWh (1992), 8% Discount Rate.
10	0.31 to 0.69

Table 2: UK Passive Solar Design Resource excluding existing "unplanned" use (see ref 3).

Environmental Aspects

PSD exploits a clean source of energy with no gaseous emissions or wastes other than those associated with the normal production of building materials and components. Because its use displaces fossil fuel energy, it also offers a reduction of CO_2 emissions which could amount to an additional 3.5 million tonnes (oxide) per year in the UK by the year 2025.

PSD has land use implications. It has been most successful in green field new towns such as Milton Keynes, which may not be repeated on such a scale.

There are also some visual considerations for PSD - such as asymmetric distribution of windows, possible reduction of privacy - although good passive solar designs can be aesthetically pleasing and lead to generally enhanced internal and external environmental conditions. All new construction and refurbishment is subject to planning and other regulatory constraints.

Economic Prospects

Unlike most other renewables, PSD does not require plant or other major capital investment. PSD measures are therefore normally assessed on their marginal costs and benefits. A simple design measure can often give energy benefits at little or no extra cost. If complex design approaches such as use of lighting controls or high performance glazing are adopted, the energy benefits must justify the additional costs on a simple payback calculation.

A particular problem in the commercial building sector is that such buildings are often not owner-occupied so that the energy benefits are not realised by the organisation paying the construction costs. There is therefore little incentive to incorporate PSD or energy efficiency measures into commercial building design unless it can be shown to increase the marketability of the building.

Opportunities and Constraints

Opportunities

- PSD in its simpler forms is an easily-applied, low-cost technology.

- The application of PSD does not depend upon major developments in the performance or cost of the technology.

- However, new developments in more complex forms of PSD currently on test are likely to yield improved energy savings in the longer term.

- Environmental awareness is growing among building designers, clients, planners and the public. This creating a desire in some professions to adopt passive solar principles to achieve pleasant, healthy and environmentally conscious buildings.

- Energy and environmental labelling schemes for buildings are becoming more widely adopted. Effective passive solar design can gain credits under such schemes.

- The passive solar programme can complement the other energy and environmental buildings programmes of Government, and by so doing will enable all programmes to meet their objectives in the most efficient and cost effective manner.

Constraints

- Lack of awareness of and expertise in PSD is widespread. Where there is such awareness there is often a lack of technical back-up and specialised guidance.

- The extent to which PSD can be used in individual buildings depends on its compatibility with the main functions of the building and on the disposition of surrounding buildings.

- There is no easy method by which the client can assess the success of the passive solar measures in a building.

- The impact of PSD is dependent on the overall level of building activity.

- Risks perceived by a conservative building industry include reduced marketability, increased construction costs and lack of examples of successful implementation.

Prospects and Categorisation

Passive Solar Design		
Occurrence	New deployment under all scenarios	
Year Contribution (TWh/y)* CO_2 savings (MtC/y)	2005 0.31 to 0.69 (heat) 0.041 to 0.091	2025 1.1 to 4.3 (heat) 0.15 to 0.57
Categorisation	Market enablement and RDD&D	

Table 3: Prospects and Categorisation.
* This contribution assumes a minimal allowance from refurbishment. If an upper estimate were used, the contribution could be increased to 5.8 TWh/year.

The Programme

Aims

1. To encourage the uptake of Passive Solar Design by:

   - assessing when the technology will become cost effective;

   - stimulating the development of the technology as appropriate;

   - removing inappropriate legislative and administrative barriers;

   - ensuring the market is fully informed.

2. To encourage internationally competitive industries to develop and utilise capabilities for the domestic and export markets, taking account of what influences business competitiveness.

3. To quantify environmental improvements and disbenefits associated with Passive Solar Design.

4. To manage the programme effectively.

Justification

Passive solar design is an easily applied technology with significant energy saving benefits. The continuing need for new build housing and non-domestic buildings provides a constant access to a growing resource which can be exploited if sufficient interest and confidence in the technology can be created in the market.

Geothermal HDR

The Technology

Description and Present Status

There is a large amount of heat just below the Earth's surface - much of it stored in low permeability rocks, such as granite. This source of geothermal heat is called "hot dry rock" (HDR). Attempts to extract the heat have been based on drilling two holes from the surface. Water is pumped down one of the boreholes, circulated through the naturally occurring, but artificially dilated, fissures present in the hot rock, and returned to the surface via the second borehole. The superheated water or steam reaching the surface can be used to generate electricity or for combined heat and power systems. The two boreholes are separated by several hundred metres in order to extract the heat over a sizeable underground volume. It is believed that a typical HDR power station would produce about 5 MW of electricity and be expected to operate for at least 20 years. In the UK, it would be necessary to drill to depths of around 6 kilometres to reach the temperatures required for electricity production (about 200°C). The success of each new installation would be sensitive to unknown local geological characteristics at that depth, adding a significant risk to the development of future installations. Experimentation in the UK has been concentrated at Rosemanowes, Cornwall. There are no commercial HDR schemes in existence anywhere in the world.

A review of the UK HDR research programme was conducted in 1990. This concluded that a satisfactory procedure for creating an underground HDR heat exchanger (or reservoir) had not been demonstrated and that electricity from commercial HDR power stations is unlikely to prove competitive in the short to medium term. Detailed examination in 1990 concluded that HDR could cost 17p/kWh at an 8% discount rate (1990 values), assuming that a commercial system could successfully be completed.

In the light of the review's conclusions, the Government decided that a new direction for the programme was necessary. The current phase of work, from 1991 to 1994, has concentrated less on research in Cornwall, and involves greater collaboration with the European HDR programme.

Despite active collaboration with the European programme, the technical difficulties of exploiting the HDR resource, identified in the 1990 review, remain. In the light of these continuing difficulties, and the likely economics of exploiting the resource, in July 1993 the Department announced the closure of the HDR programme.

DTI/DEn Programme Funding to 31/3/93	£39.4M
Number of R&D Projects to 31/12/93	49
Number of Commissioned NFFO Projects to 31/12/93	0

Table 1: Estimated programme spend (money of the day) and number of projects.

Resource

The British Geological Survey has predicted the temperatures at various depths throughout the UK. Their results for a target depth of 6 kilometres indicate that the highest temperatures are located in the granite regions of the south-west peninsula and Weardale (North Pennines). The heat in the rock can be converted to output electrical power at an overall efficiency of 3%. Therefore the Accessible HDR resource by region, omitting areas containing national parks and regardless of cost, is estimated to be 1030 TWh in south-west England and 470 TWh in Weardale. If this were to be exploited over 25 years it would result in 60 TWh/y or 7600 MW of net output at 90% availability. The practicable resource after taking account of major technical and environmental constraints, is only 500 MW. For electricity generation at under 10p/kWh (1991 prices) and a discount rate of 8% there is currently no accessible resource for HDR in the UK.

Accessible Resource (TWh/year) at less than 10p/kWh (1992), 8% Discount Rate.	Maximum Practicable Resource (TWh/year) in 2005 at less than 10p/kWh (1992), 8% Discount Rate.
0	0

Table 2: UK Geothermal HDR Resource (see ref 3).

Environmental Aspects

The main impact which an HDR generating station would have on a local community would be the need for a sizeable water supply. Local rivers in areas such as Cornwall have insufficient flow rates to support these requirements, and therefore any future development of HDR installations would ultimately depend on supplies of sea water. The technical and environmental issues involved with the use of sea water in HDR systems have not yet been studied in detail.

A number of buildings, housing power generation equipment, would mark the above ground presence of an HDR site.

The formation of the underground heat exchanger may involve the use of gels (highly viscous liquids) to force open the underground fissures and to seal short-circuit paths. If this is found to be necessary, the procedures would need to be approved by the regulatory authorities. There is little potential for atmospheric pollution at an HDR installation, although strict precautions will be needed in the storage of the organic fluids used in the electricity conversion equipment.

Economic Prospects

Because a viable underground heat exchanger has not yet been engineered, no full-scale HDR power station has been built. However, assuming a technical breakthrough in the creation of the underground heat exchanger, it is possible to estimate that a typical HDR system would cost around £40 million, with about 40% of the cost being due to the drilling operation. The annual running costs of an HDR power station would be about £1million.

The costs of generating electricity in the long term could be about 17 p/kWh and 25 p/kWh at the best UK sites, for 8 and 15% discount rates respectively, at 1991 prices. In the short term, these costs might be even higher. It is evident that the least expensive HDR-generated power would come from Cornwall where the highest subsurface temperatures are found.

Opportunities and Constraints

Opportunities

- Energy from HDR power stations is a form of non-combustion generation and has the potential to achieve savings in CO_2 emissions. Once installed, they are likely to be relatively unobtrusive and free of noise and local pollution.

Constraints

- HDR technology is not yet developed and has not been demonstrated for commercial operation.

- Estimates of the costs indicate that HDR systems will not be economically viable in the foreseeable future.

- The high cost of constructing an HDR power station and the sensitivity of its performance to the local geological characteristics render it a "high risk" project, unlikely to attract private sector funding.

- One difficulty in the development of HDR systems, and a potential limitation, is a requirement for year-round and plentiful supply of water.

Prospects and Categorisation

Geothermal HDR		
Occurrence	No deployment under the scenarios considered to 2025	
Year Contribution (TWh/y) CO_2 savings (MtC/y)	2005 0 0	2025 0 0
Categorisation	Watching Brief	

Table 3: Prospects and Categorisation

The Programme

Aims

1. Complete the existing programme.

2. Maintain a watching brief on the technology.

Justification

The 1990 Review concluded that very substantial uncertainties remain about the technical feasibility of extracting energy from HDR within the UK. Despite close participation in the CEC HDR programme there is no new evidence to change this perception. The economic judgement must be that, on present evidence, this technology is unlikely to become competitive with conventional means of generation in the foreseeable future. Continuation of the UK programme beyond existing commitments cannot be justified, given the poor prospects of the technology over the next 30 years. It is unlikely that any joint European programme over the next 5 years can change the outlook for the technology in the UK in the short to medium term and therefore the case for the UK continuing with a collaborative programme is extremely weak.

Geothermal Aquifers

The Technology

Description and Present Status

Geothermal aquifers extract heat from the Earth's crust through naturally occurring ground waters in porous rocks at depth. A borehole is drilled to access the hot water or steam, which is then passed through a heat exchanger located on the surface. If the temperature of the hot fluid exceeds about 150°C it can be used for generating electricity; otherwise it is more suited as a source of warm water. In the UK, there are very few sources with temperatures above 60C and the resource would be mainly useful for district heating systems or industrial processes.

The use of aquifers is well established in certain geologically favoured parts of the world, such as Iceland, Hungary, Italy, the USA and the Paris Basin of France. Some 6 GW of electrical generating capacity is currently installed overseas in several regions where both steam and water are produced at temperatures over 200C. In addition, many heating schemes use geothermal resources for district or process heating.

A number of test boreholes have been drilled to explore the potential in the UK; the deepest was at Larne (Northern Ireland), and others were at Cleethorpes, Southampton and Marchwood (near Southampton). The results have been generally disappointing and estimates of the available resource in the UK are small. The ownership of the borehole at Southampton, funded by Government, was transferred to Southampton City Council. The Southampton district heating scheme derives the only exploited geothermal heat in the UK from this borehole.

DTI/DEn Programme Funding to 31/3/93	£11.1M
Number of R&D Projects to 31/12/93	23
Number of Commissioned NFFO Projects to 31/12/93	0

Table 1: Estimated programme spend (money of the day) and number of projects.

Resource

In the UK the geothermal aquifer resource is found mainly in Permo-Triassic sandstone basins. Exploitation is limited to areas where the heat loads coincide with the resource, as it is not economically feasible to transport the hot fluids any significant distance. Based on a detailed assessment of the heat requirements of Grimsby, and applying the results proportionally to the other potential geothermal fields, it is unreasonable to anticipate more than a total of 50 geothermal schemes being developed in the UK. Adopting the assumption that each viable scheme would derive an annual minimum of 26 GWhth from a geothermal source, the Accessible Resource is 1,300 GWhth/year.

However, the geological conditions at the depths of interest are not known in detail and there would be considerable uncertainty about the lifetime of any proposed scheme until pumping tests had been carried out. The aquifer may be physically constrained by boundaries such as faults or less permeable strata of rock. This would cause the flow of hot water to cease after a shorter time than expected, or the temperature of the water to be below that anticipated.

Accessible Resource (TWh/year) at less than 10p/kWh (1992), 8% Discount Rate.	Maximum Practicable Resource (TWh/year) in 2005 at less than 10p/kWh (1992), 8% Discount Rate.
1.3 (heat)	0.13 (heat)

Table 2: UK Geothermal Aquifers Resource (see ref 3).

Environmental Aspects

The necessary drilling operations would have to be conducted near or within cities. The noise of such operations is a potential problem but would probably be no more than from other major construction projects. Considerable local disruption would occur during the laying of the district heating mains in existing urban areas. Once installed, there should be no disturbance except during repair or maintenance operations.

Care must be taken in the disposal of the discharge water but otherwise, aquifers are environmentally attractive, being free of gaseous emissions and having no significant problem due to visual intrusion once installed.

Economic Prospects

Geothermal heating schemes are capital intensive; about 75% of the total cost is represented by the capital and non-fuel operating costs. The drilling costs for a single borehole can represent about 20% of the total cost and there are substantial contributions from the distribution system and from operating costs. In many locations it may also be necessary to drill a second borehole to dispose of the discharged fluid (which may have a high saline content).

The cost of producing heat from the Southampton aquifer scheme, which is a single well system, including the cost of the borehole, is about 5 p/kWh using a discount rate of 8% at 1991 prices. Given that there is a significant risk, at any particular site, that a borehole may not provide the necessary hot water resource, commercial developments may require a very much higher rate of return than 8%. Geothermal aquifers based on two boreholes are less economically attractive than single well systems.

Opportunities and Constraints

Opportunities

- Aquifer schemes are environmentally friendly, being free of gaseous emissions and visually acceptable.

- The technology is well-proven and would satisfy some local energy needs.

Constraints

- The UK resource is quite small and the lifetime of each project would be sensitive to the local geological characteristics.

- Aquifer schemes are site-specific and involve a high degree of risk. Capital expenditure would be required in advance, to determine the water temperature and flow rate and potential generation of heat.

- The construction phase of an aquifer system would be noisy (like all construction projects) and care must be taken in the disposal of the discharged fluid (brine).

- Market penetration would be difficult, as any scheme would need to be assured of a ready market for heat and to be integrated into the design of local housing projects or linked with commercial or industrial projects.

Prospects and Categorisation

Geothermal Aquifers		
Occurrence	No new deployment under the scenarios considered to 2025	
Year Contribution (TWh/y) CO_2 savings (MtC/y)	2005 0 (heat) 0	2025 0 (heat) 0
Categorisation	Watching Brief	

Table 3: Prospects and Categorisation

The Programme

Aims

1. Complete the existing programme.

2. Maintain a watching brief on the technology.

Justification

The aquifer resource in the UK is limited and unlikely to be commercially exploited in the foreseeable future.

Municipal and Industrial Wastes

The Technology

Description and Present Status

The disposal of wastes produced by households, industry and commerce, along with smaller amounts of specialised wastes from industry, can pose a number of environmental problems. However, these wastes can also be used as a source of energy.

Energy can be recovered from wastes in a number of ways:

- Combustion of the waste as collected (mass-burn incineration) or after processing to reclaim recyclable components (such as metals and glass) and from the digestion of sewage sludge and organic wastes;

- By utilising the methane-rich gas produced by biological processes that occur when waste is landfilled (landfill gas collection and utilisation). The programme for landfill gas receives separate consideration in the landfill gas module.

The potential for recovering energy from wastes, and the mix of technologies that will be employed in the future, is inevitably determined by trends in waste disposal practice. There are many factors that will influence these trends, for example:

- greater implementation of waste minimisation and of recycling policies;

- developments in environmental legislation affecting both landfill practice and other disposal options;

- changes in the commercial structure of the waste disposal industry.

At present, landfilling is generally the lowest cost waste disposal option in the UK, and over 85% of all household and commercial waste is disposed of in this way. However, current trends in environmental legislation and policy indicate that waste management costs are likely to increase and the balance between alternative treatment and disposal routes may change.

Waste incineration with energy recovery is well-established overseas. The capital and operating costs of incineration plant are high and, in the UK, this makes incineration more expensive than local landfill. There are also concerns about emissions from incinerators, particularly of heavy metals and dioxins, which have resulted in the introduction of strict emission regulations. In the UK there are 30 large municipal waste incinerators, mostly built in the 1960s and 1970s. Few are fitted with energy recovery systems. All will need to be refurbished in the next few years in order to comply with EC Directives on emissions or else close.

Techniques for producing and using fuels generated from refuse have been under development for some years. Work in the UK has largely focused on the production of densified refuse derived fuel (dRDF) for use in industrial boilers. Interest is now concentrating on energy production with materials recovery in centralised resource recovery facilities (RRFs).

There is also growing interest in the development of anaerobic digestion for recovery of energy from waste. The putrescible fraction of waste is concentrated and digested to produce a methane-rich biogas. The solid residue produced may have potential as a soil conditioner or compost.

Industry also produces a diverse range of specialised wastes which require disposal and many of these have some inherent energy value if burnt as a fuel.

Very little use is presently made of these specialised wastes. This is generally because they arise in small quantities from specific operations in discrete locations and are generally regarded by industry as a nuisance to be dealt with at the lowest cost with minimum inconvenience. In the UK this usually means direct disposal to landfill through a private waste contractor. Incineration, where used, is principally adopted to cut disposal costs rather than to produce useful energy.

Currently deployed incinerator technology was developed largely in the 1960s from small scale packaged 'destructors' designed for batchwise sanitary disposal of residues. Typical plants operate at a few hundreds of kilograms per day.

Larger scale systems have also been developed overseas in response to local factors, principally the high cost of landfill disposal and stringent environmental legislation. Neither of these have been significant driving forces in the UK and currently no UK manufacturer offers such plant at scales much above one tonne per hour (3 to 5 MW thermal). However, an increasing range of technologies, including advanced rotary kiln incinerators and fluidised bed combustors are marketed by a number of internationally active equipment suppliers. Plant sizes range up to 5 tonne per hour, with multiple units being used to increase capacity above this.

The technology is currently undergoing rapid evolution, largely as a result of modern requirements for comprehensive emissions control. It is expected that standards will continue to be tightened progressively, necessitating further development of the technology and adding significantly to the cost and complexity of modern systems. The trend in utilisation is therefore expected to be towards larger scales of operation, where the economics can be advantageous - especially for wastes which can attract a substantial disposal credit.

At larger scales of operation interest is also moving towards more advanced combustion technologies. These are considered separately in the advanced conversion module.

The use of wastes as fuel is principally dealt with within the Department of Trade and Industry's Biofuels Programme, but close links have been established between this work and the Department of Environment's programme on waste management.

Since Energy Paper 55 the main activities in the current programme for waste utilisation, other than by landfill, have included the following:

- evaluation of the costs and benefits of incineration;

- evaluation and identification of promising novel approaches to incineration being tried in the UK or overseas; with feasibility studies for incineration schemes;

- evaluation, development and demonstration of systems for using densified refuse derived fuels;

- evaluation of the costs and benefits of a range of RDF processes;

- evaluation of a range of biological processing concepts under development world-wide, to identify those of relevance to the UK;

- support for the pilot scale development of high solids digestion of organic components of waste;

- evaluation, development and demonstration of technologies for combustion of specialised wastes;

- Implementation of a major strategy on environmental impacts and pollution abatement for specialised wastes;

- Dissemination of results through publications, workshops and seminars.

This work has benefited from good links with work going on internationally particularly with the IEA and the USA.

DTI/DEn Programme Funding to 31/3/93	£3.9M
Number of R&D Projects to 31/12/93	75
Number of Commissioned NFFO Projects to 31/12/93 Municipal & Industrial Waste Combustion Sewage Gas	5 projects 47 MW DNC 26 projects 33 MW DNC

Table 1: Estimated programme spend (money of the day) and number of projects.

Resource

About 25 million tonnes of household waste, 25 million tonnes of combustible general commercial and industrial wastes and 1.5 million tonnes of sewage sludge (dry basis) are generated in the UK each year. The total energy content of these municipal and industrial wastes is equivalent to around 24 million tonnes of coal per year.

The energy that could be realised depends on the technology used. The resources if all this waste were incinerated would be around 30 TWh/y. However only a portion of the industrial waste is at present dealt with alongside household waste, the remainder being dealt with by the private waste disposal sector and is unlikely to be available for energy recovery. The resource for municipal waste comprising the household waste plus the available commercial and industrial waste, is 15 TWh/y and the estimated Maximum Practicable resource is given in Table 2.

Specialised industrial wastes (such as chemicals, car fragmentiser residues, scrap tyres and hospital, meat, and wood wastes) could save energy equivalent to a further 3.5 million tonnes of coal per year. These wastes could, if incinerated, contribute a further 4.7 TWh/y (Table 2). In principle, all of the resource is Accessible, but individual arisings are often quite small and are dealt with through a multiplicity of isolated private waste disposal contracts. Commercial exploitation therefore depends on aggregating waste to sustain an economically-sized plant.

Measures introduced under the Environmental Protection Act 1990 to tighten waste management standards will lead to a significant restructuring of the disposal industry and should encourage greater aggregation of arisings.   The potential for economic energy production will depend on the delivered cost of the waste at the point of use - a function of both the waste disposal charge levied on the originator and the transportation costs associated with the aggregation process. These costs, and hence the resource potential, are influenced by other factors such as alternative disposal means (mainly landfill), pre-treatment required before transport, bulk density in transport and distance between sources of arisings.

Accessible Resource (TWh/year) less than 10p/kWh (1992), 8% Discount Rate	Maximum Practicable Resource (TWh/ year) 2005 less than 10 p/kWh (1992), 8% Discount Rate.
31.5 Municipal and General Industrial Waste including Sewage Sludge 4.7 Specialised Industrial Waste	4.2 to 5.7 MSW 0.4 Sewage Sludge 3.0 to 4.2 Specialised Industrial Waste

Table 2: UK Municipal and Industrial Wastes Resource (see ref 3).

Environmental Aspects

Using waste to replace fossil fuels can have a positive environmental impact by reducing net CO_2 emissions since much of the carbon in waste is "renewable".

Utilisation, both through incineration and anaerobic digestion, will also reduce landfill requirements and the production of methane when organic wastes are landfilled. Methane is a potent "greenhouse" gas and when emitted from landfills can also be a significant local nuisance.

Waste incinerators tend to face strong local opposition and are likely to have strict environmental and planning controls applied to them. Extensive precautions are needed to ensure that emissions are minimised and that performance meets stringent EC and national limits. In addition to meeting statutory requirements, the physical design of buildings, traffic movements, noise and nuisance mitigation require careful consideration. Sympathetic siting is required to minimise the perceived impact of the development.

Economic Prospects

The potential and economics associated with the use of municipal wastes in the future will be influenced significantly by developments in environmental legislation and trends in disposal practice. The wastes can be used to produce electricity, heat or to operate a CHP plant. The choice will depend upon the relative economics at a particular site and will be influenced by the availability of local markets for the heat.

As with municipal wastes, the potential and economics of utilising specialised industrial wastes depends largely on environmental legislation and trends in disposal practice. Niche markets will remain for the uptake of specific arisings under locally favourable conditions, particularly opportunities for onsite heating or CHP, or where there are particular constraints on alternative disposal routes, such as for clinical waste and, increasingly, fragmentiser residue, scrap tyres and certain hazardous chemical wastes. The high calorific value of these wastes, however, also make them an attractive feed stock for larger scale municipal incineration and resource recovery schemes. Consequently there is a wide range of generation costs, with the bulk of the resource being available at under 4p/kWh.

Capital costs for incineration based technologies are currently around £2.5M to £3M per MWe, but both capital and operating cost are dependent on the scale of operation, the cost of compliance with tightening environmental standards and the disposal credits available. Generation costs are typically 3 to 5 p/kWh over a 20 year project life at an 8% rate of return. However, given the high capital costs, this range is particularly sensitive to discount rate.

Opportunities and Constraints

Opportunities

- Municipal waste disposal faces increasingly stringent standards. Government policy is that, after waste minimisation and recycling, energy recovery from these wastes is a legitimate part of resource recovery and may form a part of an environmentally desirable waste disposal option.

- Under the influence of changing legislation, local authorities are required to revise comprehensively their waste management plans, taking into account long term environmental issues and overall value for money. This is encouraging consideration of a wide range of disposal options, including energy from waste.

- Environmental legislation will force many MSW incinerators to close by 1996. Prospects for energy sales may stimulate replacement of lost capacity with modern energy recovery plant.

- Electricity privatisation has opened up the market to private generation based on these wastes.

- New technologies offering the potential of higher energy conversion efficiencies (and reduced emissions) may allow the opportunity to reduce costs and improve accessible resource size.

- There is an interest in opportunities for using heat from waste for local CHP or heating schemes with higher conversion efficiencies and thus significant environmental benefits.

- Tightening landfill standards will increase costs of landfill, thus improving the competitiveness of alternative disposal routes.

- Involvement of UK industry.

Constraints

- Lack of recent UK experience of modern energy from waste or other MSW to energy schemes other than sewage sludge digestion.

- Conventional MSW to energy is currently only likely to be a cost-effective option at the most favourable sites, where alternative disposal costs are at their highest and waste is available in the largest quantities.

- There is a wide range of alternative options (including anaerobic digestion, and processing and combustion technology linked with resource recovery) which has yet to be demonstrated in the UK.

- Public perception can be hostile.

- Uncertainties about emission standards and how to achieve and show compliance with them.

- The decision making and contractual process is complex and in a state of flux.

- The costs of project preparation are considerable.

- Limited UK district heating infrastructure allowing heat sales.

Energy from small scale specialised industrial waste incineration also faces additional constraints:

- Appropriate gas cleaning systems are not yet developed and demonstrated.

- Advanced conversion technologies are not yet developed and demonstrated.

- Uncertainty over resource size, location and composition.

- Market access to the grid can be difficult for small generators.

Prospects and Categorisation

Municipal and Industrial Wastes		
Occurrence	New deployment in several scenarios in the short term	
Year Contribution (TWh/y) CO_2 savings (MtC/y)	2005 0.92 to 6.1 0.34 to 2.2	2025 0.4to 10 0.08 to 3.2
Categorisation	Market enablement via NFFO and RDD&D	

Table 3: Prospects and Categorisation.

The Programme

Aims

1. To encourage the uptake of municipal and industrial waste technologies by:

- assessing when the technologies will become cost effective;

- stimulating the development of the technologies as appropriate;

- establishing an initial market via the NFFO mechanism;

- removing inappropriate legislative and administrative barriers;

- ensuring the market is fully informed.

2. To encourage internationally competitive industries to develop and utilise capabilities for the domestic and export markets, taking account of what influences business competitiveness.

3. To quantify environmental improvements and disbenefits associated with the technologies.

4. To manage the programme effectively.

Justification

Municipal and industrial wastes currently provide some 0.8 TWh/y of electricity, principally from sewage sludge digestion and waste incineration. The technical potential of the resource is in excess of 15 TWh/y, much of which could be economically exploited using technologies well-established overseas, particularly incineration and refuse processing. The trend to larger, long-haul landfill sites, driven by stricter environmental protection standards, will increase the cost of landfill (which currently takes 90% of UK municipal solid waste), while increasing pressures for resource recovery and recycling may make energy from waste more attractive. The Programme is necessary to capitalise on opportunities and overcome barriers to the development of energy from waste. The Programme is consistent with the DTI's overall responsibility for renewable energy and complementary to the DoE's interests in recycling, waste disposal and pollution control.

Landfill Gas

The Technology

Description and Present Status

Organic wastes decay in landfills in the absence of oxygen, producing a mixture of primarily methane and carbon dioxide, known as landfill gas. Methane is a potent "greenhouse" gas and emissions from landfills can also be a significant local nuisance. However, once the environmental impacts of landfill gas are controlled, its methane content makes it a useful source of renewable energy.

Strengthening environmental protection legislation is increasing the need for effective gas control measures at landfills to protect the environment. Energy recovery can complement this prime objective of gas control.

Energy recovery from landfill gas began in the UK in the late 1970s with its use as a replacement fuel in kilns and boilers (brick kilns were a prime example). However, this approach is limited by the availability of customers sufficiently close to a landfill site to make the sale of the gas economic. Although potential heat loads still remain, power generation is now the dominant energy recovery route.

Gas turbines, dual fuel (compression ignition) engines and spark ignition engines are all used as prime movers for electricity generation from landfill gas. Engine sizes range from a few hundred kW to several MW. Conversion efficiencies range from 26% (typically for gas turbines) to 42% (for dual fuel engines). Life expectancy depends on operating conditions, but is likely to be between 10 years for the higher speed spark ignition engines and 20 years for the slower speed dual fuel engines.

Techniques for collecting and using landfill gas have been developed over the past 10 years, and are now in commercial operation. Landfill gas is collected by applying suction to perforated pipes (gas wells) sunk in the waste. The design and management of the gas collection system is influenced by specific characteristics of the site, such as depth of the waste and water table, and the primary need to control gas escape.

The gas yield depends on the amount of waste, its composition and a range of site-specific factors. Large modern landfills typically produce usable gas for over 15 years. For most applications, pre-treatment of gas is usually limited to drying and filtration, but more extensive clean-up may be required for specialised applications.

DTI/DEn Programme Funding to 31/3/93	£3.7M
Number of R&D Projects to 31/12/93	42
Number of Commissioned NFFO Projects MW DNC to 31/12/93	44 74

Table 1: Estimated programme spend (money of the day) and number of projects.

Resource

Electricity generation capacity from landfill gas had increased to 79 MWe DNC by the end of 1993, capable of producing about 0.6 TWh/y of electricity, with further savings through direct use as a fuel to heat kilns and boilers. Electricity production from landfill gas is set to increase as more schemes awarded contracts under the NFFO come on stream.

The potential for energy recovery from landfill gas depends on waste management policy and practices. About 85% of UK's municipal solid waste is landfilled at present. In the medium term, scope for improvements in landfill design and management are expected to lead to an increase in the potential for energy recovery from landfill gas. However, if waste management policy, driven by concerns over environmental impacts and resource conservation, reduces the amount of organic wastes going to landfill, then the amount of energy recovered from this source will also inevitably decline in the longer term.

Accessible Resource (TWh/year) at less than 10p/kWh (1992), 8% Discount Rate.*	Maximum Practicable Resource (TWh/year) in 2005 at less than 10p/kWh (1992), 8% Discount Rate.
5.3	6.4 to 7.7

Table 2: UK Landfill Gas Resource (see ref 3).
* The current Accessible Resource is shown; this is expected to increase in the medium term and decrease in the long term.

Environmental Aspects

Landfill sites produce gas whether it is used or not. Using the gas offers significant environmental benefits over flaring it off or simply venting it to the atmosphere. The major environmental benefits of using landfill gas are therefore those related to atmospheric emissions and the conventional energy sources it displaces.

To meet more stringent regulations on the control of landfill gas, site operators may (i) install migration barriers and gas vents; (ii) collect the gas and flare it; or (iii) collect the gas and recover energy from it. The first two options give no energy benefit, and the first also carries the risk of odours and the emission of potentially hazardous gases in trace quantities. Moreover, venting results in a significant release of methane to the atmosphere where it has a global warming potential several times that of carbon dioxide. Under the second and third options, odour problems are minimised and the methane content of landfill gas is converted into the much less harmful carbon dioxide. There will be slight differences in other gaseous emissions (e.g. oxides of nitrogen) due to the different combustion characteristics of flares and engines, but overall the net gaseous emissions from engines are unlikely to have more detrimental effects than those from flaring.

A landfill gas scheme itself will also have an impact, although this is unlikely to be significantly greater than that for equipment to control emissions for environmental protection purposes. Land needed for landfill gas power generation schemes is mainly that required for buildings to house the engine/generator sets, pumps, compressors and gas cleaning equipment. Overall plant installations (including gas extraction plant and electricity connection equipment) may be around 25 m x 25 m in area. Landscaping and sensitive design can minimise visual intrusion, which in any case is likely to be minimal because of the relatively small structures involved. Some noise is generated from the gas compressors, engine and exhaust system. With adequate insulation, siting and design, noise levels can be kept within acceptable limits. Water use is low, being limited to site services and occasional cooling water make up. Ecological impacts of landfill gas exploitation are unlikely to be significant. Vehicle movements will be few after construction. All of these impacts are minimal in comparison with those of an active or recently closed landfill site.

Economic Prospects

The example plant chosen to assess the cost of electricity generation from landfill gas is a *2* MW (electric) power station, consisting of four 500 kW spark ignition engines. The capital cost (1992 money) of £790/kW includes a proportion of the installation costs of a gas collection system, basic landfill gas clean-up (limited to water removal and filtering), the engine-generator sets, and associated electrical connection equipment. Annual costs are fixed at £130/kW and include O&M on both the gas collection and the electricity generation plant. This type of system is currently favoured by project developers with the majority of existing projects being based on this type of technology. Assuming a load factor of 88% a ten year lifetime and a 15% required rate of return the generation cost would be 3.7p/kWh.

Opportunities and Constraints

Opportunities

- Improvements in landfilling standards brought about through new environmental legislation are encouraging the installation of collection systems to control the escape of LFG. Gas collection for exploitation can complement these control measures.

- The trend towards larger landfill sites engineered for gas and leachate control for compliance with new environmental protection standards will also favour the economics of gas utilisation.

- Improvements in site management and improved performance of equipment should increase the exploitable resource of LFG and reduce costs.

- Restructuring of the waste disposal industry will remove the dependency of some projects on scarce local authority finance.

Constraints

- Although some LFG schemes are currently commercially viable without market support, for most the market price for the energy they produce is still below that needed for commercial self-sufficiency.

- Availability of finance for LFG projects is affected by the risks perceived by financiers in the performance of equipment and predictions of resource size, the relatively small financing requirements of such projects (usually below £3M) and inexperience of prospective developers in obtaining finance and meeting the lender's criteria. This reduces the utilisation of the resource.

- Some sites face difficulties in accessing the electricity market through the high costs of connection to the REC. Access to suitable heat loads limits the uptake of LFG for direct use or CHP.

- Concern over the effects of possible future environmental protection legislation, such as possible pollution abatement regulations governing emissions from LFG combustion and, in the longer term, increased recycling and incineration, acts as a further barrier to implementation.

Prospects and Categorisation

Landfill Gas		
Occurrence	New deployment in several scenarios in the short term	
Year Contribution (TWh/y) CO_2 savings (MtC/y)	2005 0.27 to 7.7 0.054 to 1.5	2025 0 to 6.3 0 to 1.3
Categorisation	Market enablement via NFFO and RDD&D	

Table 3: Prospects and Categorisation

The Programme

Aims

1. To encourage the uptake of landfill gas by:

   - assessing when the technology will become cost effective;

   - stimulating the development of the technology as appropriate;

   - establishing an initial market via the NFFO mechanism;

   - removing inappropriate legislative and administrative barriers;

   - ensuring the market is fully informed.

2. To encourage internationally competitive industries to develop and utilise capabilities for the domestic and export markets, taking account of what influences business competitiveness.

3. To quantify environmental improvements and disbenefits associated with landfill gas.

4. To manage the programme effectively.

Justification

Electricity generating capacity had increased to 77 MWe DNC by the end of 1993, capable of producing 0.5 TWh/y. Landfill gas could contribute up to 6.8 TWh/y by 2005, with further energy savings through direct use as a fuel. Landfill gas utilisation is favoured by new environmental legislation and is consistent with DoE's interests in waste management and pollution control. The Programme is also consistent with DTI's overall responsibility for renewable energy. It is necessary to overcome persisting market imperfections, for example, uncertainty over resource size, access to markets and performance of technology, which currently prevent landfill gas from achieving its economic potential. New environmental protection standards and structural changes in the waste management industry will exert a favourable influence on the development of landfill gas as an energy source.

Agricultural and Forestry Wastes

The Technology

Description and Present Status

Agricultural and forestry wastes fall into two main groups - dry combustible wastes such as forestry wastes and straw, and wet wastes such as green agricultural crop wastes and farm slurry. The first group can be combusted, gasified or pyrolysed to produce heat and/or power or a liquid fuel. The second group are best used to produce methane through the process of anaerobic digestion.

Dry Agricultural & Forestry wastes

Straw is available from cereal and other crops such as oilseeds. It is produced at crop harvest and is localised with the largest production centred in East Anglia. The use of straw as a fuel is commonplace at small scale (up to 100 kWt), using whole bale burners mainly for on-farm domestic space heating. In Denmark, there are examples of the larger scale use of straw for district heating schemes (at around 10 MWt) and power generation using CHP at around 30 MWt.

Forestry Wastes occur as the result of forestry management operations. As the value of the tree as a timber crop is in the stem, typically only that part of the tree is removed at final harvest in this country, the unwanted branches and tree tops being removed from the tree on felling and left in the forest as residues (or brash). Up to 50% of the above ground biomass can be discarded in harvesting and thinning operations and the resulting brash can present a barrier to restocking. The brash may be harvested, processed and used as fuel. Typically, the size of a project will be limited by the availability of forestry wastes in the region. Thus 5 - 15 MWe will be a typical size range.

Wet Agricultural Wastes

Animal slurries are derived from two major sources, cattle and pigs. These slurries are many times more polluting than human sewage and when not correctly managed can cause serious environmental damage, particularly by adulterating water courses and producing odours. Fines for this kind of pollution can now be as high as £20,000 per offence. One way of managing these wastes is to anaerobically digest the wastes and utilise energy in the resulting methane rich biogas. The solid fraction of the digested waste can then be used as compost and the liquid safely returned to the land. A range of digesters from 70 cu.m capacity to 1000 cu.m are commercially available. The biogas produced is used either in gas engines for electricity generation and CHP, or in gas boilers modified to accept a lower calorific value gas. These engines and boilers can be used for process or space heating. The size of the on farm units in a typical UK application will be in the 100 kWe range.

Green agricultural wastes mainly arise from the processing of root vegetables and sugar beet, as well as from the run-off from the ensiling processes and can present the farmer with a disposal problem. The use of these wastes as a fuel has not been seriously considered. However, they can be co-digested with slurry to produce methane.

Since Energy Paper 55 there has been considerable research into methods of deriving and using fuels from agricultural and forestry wastes.

- Forestry harvesting trials have been completed using the latest range of mechanisation options imported from around the world.

- As a direct result of the DTI's Wood as a Fuel programme, one pass, integrated harvesting operations have been devised for UK conditions.

- Considerable efforts have been made to investigate means of densifying straw effectively at low cost. This will give lower transport and storage costs and allow existing coal technology to be used for fuel feeding and combustion.

- Developments overseas have lead to the introduction of 'cigar burner' technology in which normal straw bales are fed into the combuster and progressively combusted.

- In collaboration with industry, a number of studies have investigated the techno-economic feasibility of constructing both straw fired and wood fired power stations under the NFFO.

- Under the NFFO a small number of commercial demonstrations of the utilisation of anaerobic digestion of farm slurries to generate biogas for electricity generation have been undertaken.

DTI/DEn Programme Funding to 31/3/93	£1.3M
Number of R&D Projects to 31/12/93	24
Number of Commissioned NFFO Projects MW DNC to 31/12/93	3 25

Table 1: Estimated programme spend (money of the day) and number of projects.

Resource

	Accessible Resource (TWh/year) at less than 10p/kWh (1992), 8% Discount Rate.	Maximum Practicable Resource (TWh/year) in 2005 at less than 10p/kWh (1992), 8% Discount Rate.
Straw	10.2	2.2
Forestry	5.0	1.4
Animal Slurries	2.9	1.4
Green farm wastes	1.0	0.3

Table 2: UK Agricultural and Forestry Wastes Resource (see ref 3).

Environmental Aspects

Agricultural and forestry waste fuels will be used as a direct replacement for fossil fuels in power generation or for industrial heating. This replacement of fossil fuels with CO_2 neutral fuels, which are also low in sulphur, will reduce emissions. Where dry wastes are to be used as fuels they will require transport from the site of production to the site of use. This reduces the energy balance associated with using these wastes as fuels and increases the emissions associated with their use. This situation can be ameliorated by siting the utilisation plant close to the point at which the fuels are produced.

Forestry wastes utilisation has several environmental impacts. Firstly, forestry sites are more amenable to early restocking if cleared and are also less likely to develop disease problems from the contaminating of new trees by brash. Moreover, the sites are more visually attractive when cleared of residues. However, the removal of whole trees from forest sites may cause soils compaction nutrient loss and water run-off which generally leads to soil erosion.

Straw burning in the field has been banned since the 1992 harvest . This may present some environmental problems associated with straw incorporation which is the most likely route for straw disposal in the absence of an energy market for this fuel. As with wood fuel, the ash left after straw fuel burning has value as a fertiliser and can be returned to the soil.

Animal slurries represent a major potential source of pollution, especially of water courses. Anaerobic digestion remains the best method of avoiding these problems with the concomitant advantage of generating a usable energy fraction.

Economic Prospects

This module covers a wide range of fuel sources, each with different conversion options, the economics of these processes vary considerably (see ref 3). Anaerobic digestion of wastes are unlikely to become economic unless driven by environmental considerations. The advanced conversion of dry wastes from electricity production show every prospect of becoming economically viable in the future.

Opportunities and Constraints

Opportunities

- Significant quantities of feedstock could be made available if a suitable supply infrastructure is developed.

- The potential to improve the efficiency and economics of using dry wastes through advanced combustion technologies is now widely recognised.

- New UK and EC environmental legislation relating to wet wastes should encourage the uptake of AD.

- Pressures for conservation of wetlands may assist the market potential for peat substitutes. Interest in the use of AD residues for such purposes is increasing.

Constraints

- Most waste fuels have a low bulk density and energy density. This has consequences for transport, storage, fuel preparation and combustion.

- Institutional barriers, such as waste supply organisation, project financing and risk management, must be overcome before larger scale systems can reach commercial exploitation.

- Advanced conversion systems have yet to be fully developed and demonstrated.

- Initial capital outlay for AD schemes is high and small businesses find their financing difficult.

- Energy production is not the driving force behind AD technology because digesters are unlikely to become profitable solely as biological energy generators.

- Currently, most farm slurries are spread on the farmers' own land. To be widely taken up, AD needs to provide the optimum waste disposal route as well as an energy production system.

Prospects and Categorisation

Agricultural and Forestry Wastes		
Occurrence	New deployment in several scenarios in the short term	
Year Contribution (TWh/y) CO_2 savings (MtC/y)	2005 0.034 to 5.2 0.0068 to 1.0	2025 0 to 6.5 0 to 1.3
Categorisation	Market enablement via NFFO and RDD&D	

Table 3: Prospects and Categorisation

The Programme

Aims

1. To encourage the uptake of agricultural and forestry wastes by:

   - assessing when the technology will become cost effective;

   - stimulating the development of the technology as appropriate;

   - establishing an initial market via the NFFO mechanism;

   - removing inappropriate legislative and administrative barriers;

   - ensuring the market is fully informed.

2. To encourage internationally competitive industries to develop and utilise capabilities for the domestic and export markets, taking account of what influences business competitiveness.

3. To quantify environmental improvements and disbenefits associated with agricultural and forestry waste technologies.

4. To manage the programme effectively.

Justification

Markets for fuels from dry agricultural and forestry wastes do not currently exist, despite these fuels becoming competitive with fossil fuels for heating in certain circumstances and all the elements to create such a market being in place. Government input is required to overcome the failure of the market to recognise the current opportunities, and bring together the disparate industrial sectors to create a market. These activities are not relying on technical developments and are therefore low risk.

With market creation will come the incentive for the potential fuel supply industry to invest in available mechanisation which will significantly lower the cost of fuel production. This, when allied to the increase in utilisation efficiency to 35% for power generation (up to a 40% increase is identified in the advanced combustion programme), will make agricultural & forestry waste fuels competitive with fossil fuels in this market.

Agricultural and, to a lesser extent, forestry wastes present similar disposal problems to municipal wastes, but do not benefit from the existence of a waste handling industry. By creating markets for 'waste to energy' schemes in the agricultural and forest products sector, such an industry will be created. An added benefit will be that pollution problems associated with poor waste management will be addressed. This is especially so with wet wastes where environmental legislation may enable energy recovery schemes.

Both agricultural and forest based industries are facing declining incomes and will welcome the opportunity to diversify into new market areas. However, Government will be required to reduce the risks at the front end of this process.

This programme presents an opportunity to stimulate markets for 'biomass' fuels in advance of energy crop production, thus stimulating the early deployment of the large energy crop resource (see Energy Crops module).

Energy Crops

The Technology

Description and Present Status

Crops which may be grown or used to produce energy in the UK range from food crops grown for energy to woody biomass. From these varied sources solid, liquid or gaseous biofuels may be derived.

Many methods for the conversion of biofuels are available, reflecting the diversity of the resource. The drier, lignin-rich materials (e.g. wood) are best converted by thermal processes: combustion, pyrolysis or gasification. Wetter biomass can be converted through anaerobic digestion to a methane-rich biogas fuel. Fermentation techniques can be used to produce liquid fuels such as ethanol.

Trials have indicated that crops of willow and poplar coppice appear the most promising option for the UK, as these crops can provide high yields and a favourable overall energy balance. Larger-scale trials are now underway that will allow current estimates of yields and costs to be confirmed. Significant improvements in yields, and reduction in production and harvesting costs, are believed to be possible.

Well-developed systems are available that can be used to provide heat or electricity from the crop at a range of scales. Conventionally electricity is produced by burning wood in a boiler, generating steam that is used in a turbine. More advanced technologies such as gasification are under development and this should allow electricity production at higher efficiency, and lead to significant reductions in costs.

A range of other energy crop systems also need to be considered - for example cereal crops to produce ethanol, or oil seed crops to produce a vegetable oil-based diesel substitute. However the estimates of potential quoted here concentrate on short rotation coppice crops.

Significant energy crop enterprises already exist in Brazil and Sweden and development programmes are underway in Scandinavia, Europe and North America. The first large-scale energy crop projects in the UK are now in progress and involve wood fuel grown as arable coppice . These projects aim to demonstrate that fuel from this source can be produced commercially and also find stable markets.

Since the EP55 review the evaluation and development of low cost techniques for producing, harvesting and using forestry residues has continued. In addition, extensive willow and poplar growing trials at various sites around the UK have been established, yielding data which has allowed the construction of the resource-cost curves presented earlier.

The first commercial demonstration of coppice production has been established through the Farm Wood Fuel & Energy Project. This project will not only show the true costs associated with growing this crop, but will also demonstrate what yields might be achieved on good arable land. In association with this project a full environmental impact assessment is being carried out to determine any environmental consequences of growing the crop.

This project has already led to coppice production costs falling through the creation of a market for cuttings as planting material and for machinery associated with the growing of this crop as a farm product. For instance, high fencing costs have been avoided by adopting farming electric fencing as opposed to forestry permanent fencing.

In the early 1980s, assessments of other arable energy crops were carried out. These concluded that such crops would be uneconomic without substantial subsidy or until the price of energy has increased markedly. Currently all energy crops are being reviewed in the light of changes since that time which might affect their viability. In addition, novel high yielding crops such as miscanthus are being assessed for the first time.

DTI/DEn Programme Funding to 31/3/93	£4.7M
Number of R&D Projects to 31/12/93	54
Number of Commissioned NFFO Projects to 31/12/93	0

Table 1: Estimated programme spend (money of the day) and number of projects.

Resource

The major restraint on the potential for developing energy crops in the UK is the availability of land. Some commentators estimate that 1 million hectares could become surplus by the year 2000, and this area could possibly rise to 5.0 million hectares by 2010. In 1993 some 630,000 hectares were taken out of production through the "rotational" set-aside scheme. It is difficult to make projections concerning future land use more than 5 years ahead and these figures are the subject of constant review.

In estimating the potential for this technology, a range of factors must be considered and some will be strongly time dependent. It is assumed that the yield of coppice rises to 21 dry tonnes/ha/y in 2010, and the conversion technologies are developed with a conversion efficiency of 25% at 1992 rising to 35% by 2010 will be achieved following the introduction of gasification.

Accessible Resource (TWh/year) at less than 10p/kWh (1992), 8% Discount Rate.	Maximum Practicable Resource (TWh/year) in 2005 at less than 10p/kWh (1992), 8% Discount Rate.
194	89

Table 2: UK Energy Crops Resource (see ref 3).

Environmental Aspects

Changes in UK land use resulting from widespread adoption of energy cropping will necessarily have a visual impact on the countryside. Local water tables may also be affected, but coppice takes fewer nutrients from the soil than conventional crops and its fibrous root structure can act as a biofilter allowing sludge to be used as a fertiliser. More diversity of flora and fauna will also accompany the change from agricultural crops to coppice. The corridors necessary for the harvesting machinery could provide up to 500,000 hectares of new wildlife habitat by the year 2010. Rural power stations are likely to be on a small-scale but will still be subject to the usual planning constraints as for other industrial development.

Use of coppice wood as fuel is environmentally beneficial, with no net CO_2 emissions from its combustion since the carbon involved is recycled. The combustion of coppice also produces emissions low in NO_x and SO_x pollutants. Under current legislation, the combustion of energy crops is considered as less polluting than industrial dry wastes and is therefore subject to less stringent pollution abatement controls.

Economic Prospects

In considering the economics of energy crop production the costs of both producing and converting the material need to be considered. These are influenced by improvements in yields and in the cost of harvesting the crop and also by the development of advanced conversion technology. The costs will also be significantly affected by factors influencing the agricultural economics, such as payments for "set-aside", wood planting , grants etc. Estimates suggest that projected increases in coppice yield and power station efficiency coupled with reductions in establishment and harvesting costs will lead to costs of electricity generation comparable with new build coal plants. The calculations for bioethanol and biodiesel are not so encouraging. They are likely to become competitive only in niche markets or with substantial subsidies.

Opportunities and Constraints

Opportunities

- Farmers and land owners are keen to identify alternative uses for land.

- The inclusion of coppice crops in Set-Aside and the Forestry Authority's Woodland Grants Scheme and improved planting regimes will reduce the high establishment costs presently imposing restraints on the technology.

- Substantial increases in crop yields from current improved arable coppice management regimes should result in medium to long-term reductions in unit costs. In addition there are opportunities to improve conversion efficiencies and costs.

Constraints

- Systems for growing energy crops have still to be demonstrated at a full commercial scale.

- The establishment of a demonstrable fuel supply chain and successful end use is essential to boost confidence in this technology.

- At today's prices, growing, harvesting, preparing and converting energy crops is not financially attractive without set-aside or Forestry Authority establishment grants.

Prospects and Categorisation

Energy Crops		
Occurrence	New deployment in some scenarios in the medium term	
Year Contribution (TWh/y) CO_2 savings (MtC/y)	2005 0 to 21 0 to 4.2	2025 0 to 150 0 to 30
Categorisation	Market enablement via NFFO and RDD&D	

Table 3: Prospects and Categorisation

The Programme

Aims

1. To encourage the uptake of energy crops by:

* assessing when the technology will become cost effective;
* stimulating the development of the technology as appropriate;
* establishing an initial market via the NFFO mechanism;
* removing inappropriate legislative and administrative barriers;
* ensuring the market is fully informed.

2. To encourage internationally competitive industries to develop and utilise capabilities for the domestic and export markets, taking account of what influences business competitiveness.

3. To quantify environmental improvements and disbenefits associated with energy crops.

4. To manage the programme effectively.

Justification

A large area of land is being removed from food production under the CAP reforms and is available for other uses such as the production of non food crops, creating the opportunity of a large potential resource size. This resource becomes competitive with fossil fuels for power generation when the cost of crop production is reduced, yields increased (using existing machinery, and well proven techniques respectively) and conversion efficiency increased to 35%, which will be achieved through the advanced combustion programme.

In certain circumstances energy crops are already competitive with fossil fuels for heating markets, especially in the rural community. In Europe, liquid fuels from energy crops are already in production, and fiscal changes to support their production are being proposed through the EC.

Combustible energy crops will not require price support as applied to food crops, but initially an establishment grant may be necessary. This grant may also be desirable as a means by which the programme can control clonal selection and plantation design. Such a grant is already available for coppice simultaneously reducing support for the farming industry and overcoming the problems and cost of food over supply.

All the elements required to grow and utilise energy crops are available, yet no market for these products exist. This is due to both the disparate nature of the farming and fuel utilisation sectors leading to a failure of the market to recognise the opportunities, and also the lack of commercial demonstration of the technology.

Adopting energy crops can have considerable environmental benefit through both being 'low input' in nature and providing increased habitat biodiversity in the countryside, as well as providing a renewable fuel, which is cleaner than many conventional fossil fuel sources.

This programme will augment the work proposed in the agricultural and forestry wastes programme by increasing the size of the resource available to markets created for 'biomass' wastes.

Advanced Conversion

The Technology

Description and Present Status

Three principal advanced conversion technologies are considered here: pyrolysis, gasification and liquefaction. These thermochemical processes produce solid, liquid and gaseous intermediate products from biofuels which can be burned to generate heat and/or electricity, or used to simulate fossil fuels. These 'intermediates' as they are called, can be combusted directly in an engine or turbine to produce power avoiding the steam cycle. This gives a substantial increase in conversion efficiency at the scales relevant to biofuels with capital and operating costs comparable with conventional plant. The economic viability of electricity production from energy crops, forestry wastes and straw is thus significantly improved.

These processes promise to be inherently less polluting than conventional incineration, the main technology used for producing electricity from municipal and industrial wastes. By reducing the costs of pollution abatement and offering a more secure disposal route which satisfies stringent pollution legislation these advanced technologies will enhance the prospects of energy-from-waste schemes. For sewage sludge, net power export by conventional incineration is difficult to achieve. Gasification may prove to be the best option for both disposal and power generation.

Pyrolysis is the thermal degradation of a feedstock in the absence of an oxidising agent to produce gas, liquid and char. The pyrolysis oil may be used to fuel internal combustion engines or it may have value as a chemical feedstock. By this route, it may be possible to convert whole crops including the straw for the generation of power or production of liquid biofuels.

Gasification is the substoichiometric oxidation (i.e. less oxygen than that required for complete combustion) of a feedstock to produce a gaseous mixture of carbon monoxide, carbon dioxide, methane and hydrogen.

Gasification equipment is now available in sizes which allows electricity to be generated in units ranging from 100 kWe to 30 MWe using biomass. This covers the range of interest for crops, forestry wastes and straw. Power production from waste is more problematic but it is approaching the pilot scale demonstration stage. Within all major national biofuels programmes, gasification has been rising in prominence with a number of countries engaged in large scale demonstrations of the technology.

Liquefaction is a relatively low temperature process which operates in the liquid phase with high pressure hydrogen injected to liquefy the feedstock. There are a number of technical and engineering requirements which are likely to prevent this process being economically viable.

Advanced Conversion is a new element introduced into the renewables programme following analyses of the cost of generating electricity from energy crops. The opportunity to increase the conversion efficiency presented by advanced conversion offers greater generating cost reductions than are likely to be achieved by reducing fuel costs alone. The municipal and industrial wastes programmes have highlighted the impact of pollution legislation on the economic viability of schemes. Advanced conversion has risen in importance because of its long term potential to offer a lower cost route to compliance with ever more stringent legislation.

Since 1977, a number of R&D projects and studies have been supported on the fringe of the other biofuels areas investigating, for example, methanol production from wood, gasification of RDF and liquefaction of wastes. It is now clear that the work should be given an individual identity and pushed forward with close UK industrial involvement.

DTI/DEn Programme Funding to 31/3/93	£1.1M
Number of R&D Projects to 31/12/93	27
Number of Commissioned NFFO Projects to 31/12/93	0

Table 1: Estimated programme spend (money of the day) and number of projects.

Resource

The resources are described under the relevant biofuels modules - Municipal Wastes, Industrial Wastes, Agricultural Wastes and Crops. The scope for advanced conversion processes cannot be assessed properly until sufficient information becomes available on the improvements in efficiency achievable and on the capital and operating costs associated with them.

Environmental Aspects

Work on advanced conversion is at too early a stage to be certain of the environmental impacts of the technology. However, for a number of reasons the impact is expected to be lower than that of conventional steam plant.

Improved conversion efficiency means that the carbon dioxide emissions per unit of electricity generated is lower. The conversion efficiency is expected to increase from 7%, for steam plant, to 25%, for gasification, at the 100 kWe scale and from 20% to 35% at the 15 MWe scale: carbon dioxide emissions will, therefore, be reduced by over 70% and by over 40% at the two scales respectively. Emissions from other pollutants are likely to be lower for the following reasons.

- The heavier pollutants will be retained in the ash because of the relatively low reactor temperatures.

- The volatile pollutants, hydrogen chloride and SO_x, can be removed from the gas output before combustion, giving an advantage over conventional plant where it is necessary to scrub the much larger flue gas flow.

- It is possible that the combustion process will be easier to optimise and so the carbon monoxide and NO_x emissions will be reduced.

- Preliminary work has indicated that the ash from gasification has good leaching properties allowing safer disposal in landfills.

- As a steam cycle is not used, there is no need for water abstraction and disposal, nor for cooling towers which can cause visual intrusion.

The location of plant which utilises advanced conversion will be the same as for conventional plant. The plant life expectancy is expected to be similar and decommissioning should not present any special problems.

Economic Prospects

At this time it is not possible to present a detailed discounted cash flow analysis of an advanced conversion plant. However the following statements can be made, based on current knowledge.

- There is an international consensus that gasification could offer the most cost effective means of utilising biofuels.

- The conversion efficiency gains are widely accepted.

- Data available on prospective projects shows capital costs comparable with conventional steam plant.

- Operating costs are unlikely to differ greatly from the costs for steam plant.

The gasification route for biofuel conversion, therefore, offers prospects of a significant increase in conversion efficiency with commensurate reductions in electricity generating costs.

Opportunities and Constraints

Opportunities

- Involvement in international activities under the IEA Bioenergy Agreement and within the CEC's JOULE and THERMIE programmes provide excellent opportunities for information exchange and the generation of collaborative projects.

- Interest in liquid biofuels is continues in Europe with the likelihood of EC programmes and funding opportunities arising in the future.

- As the importance of biofuels for power generation is realised, so international export markets for Advanced Conversion equipment will arise.

- In comparison with conventional steam cycle electricity generation, there is scope for a conversion efficiency improvements and therefore, assuming a 50% uptake of Advanced Conversion Technology by 2025, the biofuels resource may be considered to be increased by 30% or more.

Constraints

- Limited UK technical knowledge to support development activities.

- Insufficient information to optimise the match between technologies, feedstocks and scales of operation.

- Capital costs are seen to be high for demonstration plant in comparison with "off-the-shelf" conventional plant.

- Financial institutions unwilling to invest in novel technology.

Prospects and Categorisation

Advanced Conversion		
Occurrence	New deployment in some scenarios in the medium term	
Year Contribution (TWh/y)[1] CO_2 savings (MtC/y)	2005 0 to 24 0 to 4.8	2025 0 to 150 0 to 30
Categorisation	Market enablement via NFFO and RDD&D	

Table 2: Prospects and Categorisation.
[1] *Potential contribution from agricultural & forestry wastes and energy crops via gasification technology.*

The Programme

Aims

1. To encourage the uptake of advanced conversion technologies by:

- assessing when the technology will become cost effective;
- stimulating the development of the technology as appropriate;
- establishing an initial market via the NFFO mechanism;
- removing inappropriate legislative and administrative barriers;
- ensuring the market is fully informed.

2. To encourage internationally competitive industries to develop and utilise capabilities for the domestic and export markets, taking account of what influences business competitiveness.

3. To quantify environmental improvements and disbenefits associated with advanced conversion technologies.

4. To manage the programme effectively.

Justification

The economic use of energy crops and small scale power systems benefit greatly from increases in the efficiency of conversion to electricity. One of the major barriers to the use of municipal and industrial wastes as fuels is economic compliance with emissions legislation. Advanced Conversion technologies have the potential to overcome both of these obstacles, but support is needed to develop and demonstrate commercial systems. These technologies are also highly relevant to the production and upgrading of liquid biofuels derived from crops.

Advanced Fuel Cells

The Technology

Description and Present Status

Fuel cells are electrochemical devices which convert the energy of a chemical reaction directly into electricity and heat. They are similar in principle to primary batteries except that the fuel and oxidant are stored externally, enabling them to continue operating as long as reactants are supplied. The fuel cell is a potential alternative to conventional energy conversion technologies offering prospects for high efficiency and low environmental emissions. Fuel cells could find future applications in Combined Heat and Power and transport.

Each cell consists of an electrolyte sandwiched between two electrodes. Fuel is oxidised at the anode, liberating electrons which flow via an external circuit to the cathode. The circuit is completed by a flow of ions across the electrolyte that separates the fuel and oxidant streams. Practical cells typically generate a voltage of around 0.7 to 0.8 volts and power outputs of a few tens or hundreds of Watts. Cells are therefore assembled in modules known as stacks and connected electrically in both series and parallel to provide a larger voltage and current. Fuel cells are usually classified by their electrolyte.

Alkaline fuel cells (AFC) are relatively simple devices and were the first to be developed. An alkaline electrolyte, such as potassium hydroxide, is used with activated nickel or precious metal electrodes. The AFC is usually limited to operation with pure hydrogen and oxygen and it is likely to become more important if hydrogen becomes more widely available. Although it can now be considered a developed technology there are still significant programmes, aimed at urban transport applications, particularly in Belgium (Elenco).

Phosphoric Acid Fuel Cells (PAFC) employ a phosphoric acid electrolyte and platinum or platinum-ruthenium electrodes. A number of demonstration units have operated for several thousand hours in Japan and the USA and the largest installation to date has an output of 11MW.

PAFC systems are expected to play a significant role in Japan, where electricity costs are high and dispersed generation is preferred. Opportunities in Europe are expected to be more limited. In the longer term more advanced fuel cells, currently at an earlier stage of development, are expected to replace the PAFC in most applications.

Solid Polymer Fuel Cells (SPFC) use a sulphonic acid electrolyte which is incorporated into a polymer membrane so that the electrolyte is effectively solid. They operate at 80C and they use platinum-based catalysts. The SPFC can be run directly on hydrogen or on reformed methanol or natural gas but as platinum is poisoned by carbon monoxide this must be removed during fuel processing.

Although SPFCs are used in specialist space and military applications they remain relatively undeveloped and are not, as yet, commercially viable for wider use. They are currently manufactured for interested bodies, although not on a commercial scale and prices are high due to the cost of materials, manufacturing costs and the need to recover R&D costs.

Molten Carbonate Fuel Cells (MCFC) use an electrolyte consisting of an alkaline mixture of lithium and potassium carbonates, which is liquid at the operating temperature of 650C, and is supported by a ceramic matrix.

The operation of an MCFC is fundamentally different to that of other fuel cells and it is tolerant to both carbon monoxide and carbon dioxide. Hydrocarbon fuels, including coal gas, may be reformed directly at the anode.

Development programmes in Japan and the USA have produced many small prototype units in the 5 to 20 kW range. Materials problems however remain an issue and still limit the useful lifetimes of the stacks.

Solid Oxide Fuel Cells (SOFC) use a solid zirconia based electrolyte which conducts over the 850C to 1000C temperature range.

Natural gas is the preferred fuel for most applications and may be reformed within the cell, in an external reformer or by using a combination of both. Other fuels, such as coal gas could be used but, although the SOFC is thought to be more tolerant to sulphur than other types of fuel cell, some clean-up during fuel processing would still be required.

To date the Programme has undertaken R&D to evaluate SPFC and SOFC. On SPFC, research has addressed improved electrode designs, increased catalyst tolerance to carbon monoxide poisoning, improved fuel processors and gas clean-up systems. On SOFC, work has focused on materials, fabrication processes, cell and stack development and system integration.

DTI/DEn Programme Funding to 31/3/93	£0.59M
Number of R&D Projects to 31/12/93	28

Table 1: Estimated programme spend (money of the day) and number of projects.

Resource

In principle any substance capable of chemical oxidation can be used as fuel but hydrogen is preferred due to its high reactivity. This can be produced from hydrocarbon fuels such as methanol, natural gas or coal gas by a process known as reforming. With high temperature fuel cells, it may be possible to increase efficiency and simplify the system by reforming fuel within the stack. Low temperature systems will require the fuel to be reformed externally. Alternatively, hydrogen may be produced by the electrolysis of water, possibly using renewable energy sources.

Environmental Aspects

With their potential for high efficiency at both full and part load, fuel cells offer the prospect of low carbon dioxide emissions. They also offer low emissions of other atmospheric pollutants, such as oxides of nitrogen and sulphur, carbon monoxide, particulates and hydrocarbons.

These characteristics coupled with modular construction and quiet operation could make fuel cell systems particularly attractive for small scale distributed power generation and combined heat and power (CHP) systems. They may also offer a less polluting alternative to the internal combustion engine in transport and in the longer term have applications in large scale power generation.

Economic Prospects

As advanced fuel cell technology is still in the early stages of development, it is not possible to provide detailed information on generation costs at this stage in the programme. However, fuel cells show promise for being competitive in combined heat and power systems in the medium to long term, and may offer an environmentally attractive option for transport applications.

Opportunities and Constraints

Opportunities

- The fuel cell is intrinsically a very attractive energy conversion technology, offering a modular construction to suit a wide range of capacities, quiet operation, low environmental impact and high efficiency.

- Developments in materials technology and the UK's world class skill base in this area offer new opportunities to take forward the development of advanced fuel cell technology.

- Regulatory, legislative and commercial developments, together with increasing public awareness of environmental issues in the UK and overseas, point towards enhanced market opportunities for advanced fuel cell systems.

- Participation in complementary programmes elsewhere can be exploited to gain access to information at low cost, to gear up the UK's efforts and to influence the direction of international programmes towards the UK's interests.

- The programme will build on earlier work supported by Government and the emerging fuel cell community that it has fostered.

Constraints

- High materials and fabrication costs.

- Inadequately defined application requirements and development priorities.

- A gap between current performance and application requirements.

- A lack of a natural industrial base to provide a focus for advanced fuel cell technology in the UK.

- The potential users' lack of experience of, and confidence in, the technology.

- The need to establish a significant market in order to reduce costs through mass production.

Prospects and Categorisation

Decentralised Power Generation and Transport		
Occurrence	Insufficient data to allow modelling	
Year Contribution (TWh/y) CO_2 savings (MtC/y)	2005 unknown unknown	2025 unknown unknown
Categorisation	Assessment and RDD&D	

Centralised Power Generation		
Occurrence	No deployment under the scenarios considered to 2025	
Year Contribution (TWh/y) CO_2 savings (MtC/y)	2005 0 0	2025 0 0
Categorisation	Watching Brief	

Table 2: Prospects and Categorisation

The Programme

Aims

1. To provide Government with an evaluation of the technical and economic prospects for advanced fuel cell systems.

2. To provide a basis on which industry can assess the prospects for the commercial development of advanced fuel cell systems and to encourage UK industry to develop capabilities to supply the domestic and export markets.

3. To stimulate the development and deployment in the UK of economically attractive advanced fuel cell systems.

Justification

The fuel cell is a potential alternative to conventional energy conversion technologies offering prospects for high efficiency and low environmental impact. Their modular nature, quiet operation and high conversion efficiency could make fuel cell systems particularly attractive for Combined Heat and Power (CHP), distributed power generation and transport applications. Although the technology is promising and there is a major world-wide effort to develop attractive systems, further research and development is needed before the potential of, and prospects for, advanced fuel cells can be evaluated fully. A structured programme will enable both the DTI and industry to gain an informed view of the technology's potential and prospects at reduced cost, whilst benefiting from a wide range of international activities. In addition it will help to establish a UK manufacturing capability and the necessary infrastructure to bring the technology to market.

The SOFC and the SPFC are two of the most promising advanced fuel cell systems, but both are at an early stage of research and development. For this reason and due to the level of industrial interest in these systems from UK organisations, the UK programme focuses mainly on SPFC and SOFC technologies.

Commercialisation

Background

Description and Present Status

With the privatisation of the electricity supply industry in the UK over the period 1990-93, the Government has introduced statutory obligations on the industry to purchase power from renewable energy schemes. The DTI's programme has adapted by shifting emphasis towards the more commercially promising technologies and concentrating on deployment issues. In 1991 the Renewable Energy Commercialisation Programme was created to address the cross-technology issues and non-technical barriers which could impede the commercial deployment of renewables. Many of these result from the lack of experience with renewable energy of those involved with deployment - planners, financiers, lawyers, regulators and even developers themselves. The programme works with the technology programmes to identify and tackle the generic non-technical barriers, provides expertise in these areas and co-ordinates the relevant activities across the DTI renewables programme.

The programme can best be described through its main components:

Planning - planners faced with planning applications for renewable energy schemes have had to make decisions in the absence of local policies on renewable energy, and often without adequate knowledge of local renewables resources or the environmental impact and benefits of relevant technologies. The programme is active in helping to formulate national planning guidance for renewable energy (e.g. Planning Policy Guidance Note No.22 in England and Wales, issued by DoE in February 1993) and assists local authorities in assessing the planning and environmental implications of renewables deployment.

Environment - renewables can bring substantial global environmental benefits by reducing our use of conventional fuel sources. However this must be balanced against their predominantly local environmental impacts. The programme is seeking to quantify the benefits and impacts so that a more rigorous comparison with conventional fuels can be made. This involves the development of suitable techniques and liaison with similar activities in other countries.

Financing - renewables are capital intensive investments and must compete with a variety of other energy technologies. Investors and lenders will want to be confident that returns are commensurate with risks and that a workable commercial framework exists. The programme seeks to assist the deployment of renewable energy by involving major financial institutions, identifying means of increasing the availability and reducing the cost of project finance, producing suitable guidance for developers and financiers and, where possible, by easing contractual constraints. Whilst deployment in the electricity sector is currently encouraged through a premium market-place, the programme aims to identify other cost-effective mechanisms to assist the integration of renewables into the energy supply market.

Non Fossil Fuel Obligation - the NFFO has been very successful in stimulating an initial market-place for certain renewables. The programme assists the DTI to discharge its functions with respect to NFFO by co-ordinating ETSU's advice and monitoring of NFFO projects.

Resource Studies - whilst it is recognised that a substantial renewable energy resource exists, the proportion of this that can be realistically exploited in the near to medium term is constrained by many factors - planning, integration with the supply system and, not least, cost. The programme has set up joint resource studies with companies interested in assessing the exploitable resource at a local level. Information is published in the form of summary reports to stimulate local, regional and national debate about renewables deployment issues. By mid-1993 nine such studies were ongoing.

Renewable Energy Skills Resource - deployment has accelerated dramatically since the introduction of NFFO and, in some cases, is being impeded by a shortage of suitably qualified personnel to assess, implement and enable projects. The programme aims to assess the extent of the problem and propose whether there is a role for the Government renewables programme in alleviating it. This is initially being achieved through the monitoring of NFFO projects.

Other topics covered under the Commercialisation programme include:

- the integration of electricity from renewable sources into the electricity supply infrastructure. The current distribution network is generally designed for large centralised power stations and integration of large numbers of small, geographically dispersed generators poses a series of economic, technical and institutional problems.

- NFFO is stimulating the creation of an active renewables supply industry in the UK, ranging from equipment supply to a wide range of deployment services (consultancy, planning, legal, monitoring, etc.). The programme's role is to assess the opportunities for this industry to exploit these skills outside the UK and, where possible, to assist the industry to access this export potential.

DTI/DEn Programme Funding to 31/3/93	£1.6M
Number of R&D Projects to 31/12/93	24

Table 1: Estimated programme spend (money of the day) and number of projects.
*This figure includes work at ETSU and contracted out.

Opportunities and Constraints

Opportunities

- The opportunity to establish a commercial framework suitable for new and renewable energy technology deployment, including the integration of renewable energy into the UK energy supply infrastructure.

- The potential to enable local planning authorities develop policies for renewable energy deployment

- The potential to improve our understanding of the environmental implications of renewable energy in order to assist planners and policy makers in evaluating the benefits and impacts.

- The scope for attracting new interest and investment into new and renewable energy technologies through targeted cost-shared business opportunity studies

- The potential to maximise the cost effectiveness of market incentives such as NFFO

- The opportunity to enhance the successful implementation of projects by providing developers with suitable guidance

- The potential to assist the UK renewables industry identify and exploit export opportunities.

Constraints

- An energy supply infrastructure and a commercial, regulatory and institutional environment often ill-suited to new and renewable energy technologies

- Unfamiliarity of planning authorities with renewable energy deployment issues

- Poor knowledge of commercially viable opportunities for industry and commerce

- Difficulties in raising investment finance for new and renewable energy projects

- A shortage of suitably skilled personnel to assess and implement new and renewable energy projects

- Insufficient experience of commercial deployment of new and renewable energy technologies amongst prospective developers.

The Programme

Aims

1. To facilitate the successful commercial deployment of new and renewable energy technologies in the UK.
2. To assist the transfer of UK new and renewables expertise to overseas markets.

Justification

The Non Fossil Fuel Obligation has provided a major incentive to the deployment of renewable energy technologies, but has also highlighted a number of significant commercial barriers to widespread exploitation. These include unfamiliarity on the part of local authorities with renewable energy, difficulties for schemes in negotiating both investment finance and required contracts and the general lack of experience of those dealing with schemes - developers, planners, bankers and regulators. These difficulties are to be expected for technologies new to the commercial marketplace, particularly as new and renewable technologies are generally exploited at a much smaller scale than conventional energy supply technologies. However if the technologies are going to achieve their market potential, there is a need for Government to address these barriers, with the aim of creating an appropriate institutional and commercial framework.

The Commercialisation Programme aims to meet these needs on a cross-market, cross-technology basis by working with the technology and marketing programmes to identify the barriers, fully understand their implications and propose appropriate solutions. By developing expertise on these generic issues, the programme provides a necessary complement to the technology and marketing and promotion programmes. It also acts as a focal point for a series of regional studies with businesses and planning authorities, aimed at stimulating a positive regional framework for deployment, and encouraging industry to incorporate new and renewable energy technologies into their company business plans. By co-ordinating the relevant activities of the technology teams and by acting as focal point for the DTI and other parties on these issues, the Commercialisation Programme enhances the effectiveness and efficiency of the overall programme.

Marketing

Background

Description and Present Status

The targeted dissemination of results from the Technology and Commercialisation Programmes and the NFFO is an essential element in stimulating the development and deployment of new and renewable energy technologies; the Marketing Programme provides the main vehicle for information dissemination activities. Underpinning the programme is market research, to increase market understanding and knowledge and to provide feedback on the impact and effectiveness of specific marketing activities.

The Marketing Programme seeks to ensure that:

- the information needs of each of the different markets are appropriately and fully met

- an appropriate balance is struck between the complementary flows of information from the different Technology Programmes, the Commercialisation Programme and the NFFO

- those marketing tools which maximise the impact of the disseminated information are utilised.

The Marketing Programme also provides a focus for:

- market research activities

- targeted marketing activities, including publications, events and exhibitions

- general awareness promotions as well as enquiries processing and contact database management.

Established in 1987 to provide a mechanism for technology transfer activities in support of the Technology Programmes the Marketing Programme was comprehensively reviewed during 1991/92 to take account of the following issues:

- Ensuing deployment of renewables technologies, brought about by the NFFO

- The removal of key institutional barriers, e.g. waste abstraction charges

- Changes in agricultural practice

- Heightened environmental interest and awareness.

- Limited industrial and commercial understanding of new and renewable energy technologies

- Lack of commercial information

- Sparse market knowledge

- Limited guidance for developers, planners and financiers

- Adverse public perceptions of certain technologies, e.g. wind farms and waste incinerators.

The outcome of the review was a reorientation of the Marketing Programme to the current market sector-based approach, whereby an appropriate marketing mix is adopted to address the specific information needs, the opportunities and the barriers to investment in new and renewables technologies within each sector.

Activities under the Marketing Programme are targeted on five market sectors:

- Utilities (electricity, heat, fuel and water supply companies)[1];

- Local Government and Waste Management (local planning, health, education and waste regulation authorities; LAWDCs and private waste management companies)[1];

- Agriculture & Forestry (farmers, landowners, land agents, forestry industry)[1];

- Industry & Commerce (process industries, e.g. bricks and ceramics, food and drink etc.; financial institutions)[1];

- Influencers (general public, pressure groups, students).

[1] *To all these "developer" sectors can be added the following "enablers": equipment manufacturers and suppliers; trade associations and professional institutions; consultants and research organisations; regulatory bodies; and professional/technical media.*

In addition, work is supported in two other areas, international liaison and non-market specific activities.

Since the inception of the Marketing Programme the number of industrial and commercial enquiries for information has risen from 800 in 1989 to nearly 3000 in 1993; the proportion of the total number of enquiries coming from industry and commerce now stands at 30%. The number of delegates at DTI events has also risen from just over 1400 to 3000. Independent surveys have shown that in 1991 31% of industry and commerce thought that renewables are relevant to them compared to 25% just two years earlier. More significantly, 79% believe that renewables will become increasingly relevant to them. Indeed, further independent research in the utilities sector in 1992 has shown that 82% believe renewables to be relevant to them now, with two-thirds having received marketing information - showing the level of penetration achieved.

The number of enquiries from the general public and students has also increased significantly, from around 3200 in 1989 to nearly 7000 in 1993. Independent research has shown that the level of awareness of different renewables technologies increased from 27% in 1988 to 39% in 1991. Evidence of understanding is shown by the primary benefit associated with renewables changing - from cost savings (down from 35% to 24%) to environmental benigness (up from 14% to 33%).

Whilst the Marketing Programme cannot claim sole responsibility for these changes in levels of awareness and understanding, it most certainly will have contributed to them through the types of activities employed. Further verification is provided by the fact that nearly 90% of the successful applications under the first and second NFFO Orders have received Marketing Programme literature or had contact with ETSU staff.

DTI/DEn Programme Funding to 31/3/93	£7.2M*
Number of Projects to 31/12/93	132

Table 1: Estimated programme spend (money of the day) and number of projects.
*Includes work at ETSU and contracted out.

Opportunities and Constraints

Opportunities

- A number of new and renewable energy technologies have matured to the point where they are being deployed commercially in the UK (albeit mostly with support from NFFO) - agricultural and industrial waste combustion, landfill gas, wind, passive solar design and small scale hydro. Further take-up of these technologies can only be facilitated via dissemination of comprehensive and independent information to potential replicators.

- The NFFO currently is the primary driving force for deployment of renewables. The NFFO Orders have created a considerable market demand, and thus far have been oversubscribed. NFFO projects are generating a very considerable body of technical, economic and environmental data that has to be effectively used to ensure successful future installations which adopt best practice. Through this, the credibility of new and renewable energy technologies can be properly demonstrated and the longer-term (i.e. post-NFFO) widespread deployment assured.

- Many institutional barriers have come down - e.g. the removal of abstraction charges for hydro projects, privatisation of the electricity and water industries, exemption of small scale biomass combustion plant from pollution abatement regulations, the creation of LAWDCs. These give enhanced opportunities for the technologies, again which can only be exploited through a programme of information dissemination.

- Changes in agricultural practice because of reducing CAP support are giving rise to opportunities for replacement of traditional crops by arable coppice. Similarly urban forestry initiatives may lead to fuelwood production. The ban on straw burning in the fields after 1992 may encourage use of straw as fuel at farms and in rural-industry. All these opportunities should be exploited through a rational programme of information dissemination based on existing RD&D projects in these areas.

- The heightened environmental interest and awareness among all sectors of the community has created a real surge in demand for much more closely targeted information on renewables.

Constraints

A number of barriers to the take-up of renewables have also emerged, which are being tackled through the Technology and Commercialisation programmes and the NFFO. The Marketing Programme is the vehicle for information dissemination to overcome these:

- Decision-makers in industry and commerce often are not aware or have limited understanding of new and renewable energy technologies. The technologies may be viewed as irrelevant within certain market sectors.

- New and renewable energy technologies are not always fully proven and their environmental impact may be inadequately understood by industry and commerce.

- There is limited information on the commercial performance of the technologies.

- The weak financial position of some market sectors inhibits investment in the technologies and the financial institutions who could invest on their behalf may not recognise their merits.

- Market knowledge is sparse, with the result that potential equipment suppliers may not consider entering the market and potential users do not know who to approach for information or support. Consequently the UK new and renewable energy equipment supply industry is weak.

- Developers often lack experience in implementing real projects, and little overall guidance is available to them.

- Waste Disposal Authorities, LAWDCs and private waste disposal companies may not be fully aware of all of the options for waste recycling or disposal.

- Adverse public perceptions of, in particular, wind farms and waste incinerators can hinder their development, and may be based on poor understanding or mis-information.

The Programme

Aims

1. To develop a full understanding and knowledge of key industrial and commercial market sectors where new and renewable energy technologies are likely to be deployed in the short to medium-term and, in conjunction with the Technology and Commercialisation Programmes, to identify specific investment opportunities.

2. To increase industrial and commercial awareness, understanding and knowledge of new and renewable energy technologies, and thereby establish their credibility as potential energy sources.

3. To demonstrate the relevance of the technologies to target industrial and commercial enterprises, and thereby:

   • encourage investment in the RD&D Programme so that technologies may be developed appropriate to the market;

   • inform investment under the NFFO, thus facilitating adoption of best practice and securing successful initial deployment; this will create the conditions for wider deployment post-NFFO as the technologies become economically viable;

   • inform investment in non-Grid connected projects, thus facilitating best practice and securing successful deployment wherever they can be economically utilised for this application in the short to medium-term.

4. To provide timely, accurate, comprehensive and independent information to all sectors.

5. To measure and assure the impact, effectiveness and value-for-money of the Marketing Programme, utilising this feedback in developing future plans.

Justification

The primary aim of the New and Renewable Energy Programme is the development of a self-sustaining market for each of the technologies as they become technically, economically and environmentally viable. In order to achieve this, information generated from the Technology and Commercialisation Programmes and through the NFFO needs to be transferred to industry, commerce, the public and students in a targeted and cost-effective manner; the Marketing Programme provides the main vehicle for information dissemination activities. By raising awareness and understanding, and by providing comprehensive, independent and objective technical, economic and environmental information to target industrial and commercial sectors, the Marketing Programme stimulates development and deployment of new and renewable energy technologies. Without the Marketing Programme, the New and Renewable Energy Programme could not fully meet its aims.

Through its market sector-based approach, the Marketing Programme has been specifically designed to address opportunities for, and barriers to, deployment of new and renewable energy technologies and thereby support the commercial take-up of renewables in the short-term (under the NFFO) and in the medium and longer-term outside and beyond the NFFO.

The Programme provides the means for delivering results from over £250 million of Government support for RD&D and nearly £1 billion of industrial/commercial investment in NFFO thus far.

Notes

Resource estimates *(table 2 in each module)*

Two terms are used to describe the resource size of appropriate new and renewable energy technologies: the Accessible Resource and the Maximum Practicable Resource. These need some explanation to allow correct interpretation.

The Accessible Resource represents the resource which would be available for exploitation by a mature technology after only primary constraints are considered. For example; for wind power National Parks and physical constraints, such as housing, roads and lakes are excluded from the calculation of the resource size. For most technologies this measure of resource is still large and its full exploitation unlikely to be acceptable as it would result in power plant in every available location. Moreover the cost ceiling of 10p/kWh is high compared with the current pool price (less than 3p/kWh on average). Whilst the Accessible Resource may therefore indicate a considerable theoretical potential for renewable energy, it is not a realistic measure of the actual contribution which may be made in future. However the Accessible Resource provides a starting point from which realistic estimates of the likely resource size can be made.

To assess the contribution that technologies might make in the real world a more realistic measure, taking account of additional constraints upon their deployment, is required. This measure is here called the Maximum Practicable Resource. In deriving the estimates of the Maximum Practicable Resource that are presented in this Annex an examination of the constraints on the deployment of each technology, and how they might change with time, was undertaken. Many of these constraints - regulatory, sociological, environmental - are not susceptible to objective scientific assessment and subjective judgements were often required. These judgements were informed by studies undertaken as part of the earlier programme and existing experience of deployment which in most cases is extremely limited. Moreover for each technology the UK as a whole was assessed without detailed analyses of the complex system integration and operational issues at a regional level, but ongoing regional assessments have suggested that these issues could significantly constrain the Practicable Resource. The resource estimates presented here are therefore for the "Maximum" Practicable Resource.

The Accessible Resource and the Maximum Practicable Resource provide estimates of the total deployment of a technology under certain circumstances, not just estimates of the new deployment, existing schemes are therefore included in the estimates where appropriate. An explanation of the derivation of the resource estimates for each technology is given within the individual modules in the Assessment of Renewable Energy for the UK (Ref 3).

Number of scenarios under which new uptake occurs *(table 3 in each module)*

Six scenarios and three discount regimes were used to assess the possible future contribution from each of the appropriate technologies: a total of 18 cases. This Annex presents the number of cases in which new deployment was estimated and the timing of this new deployment. A brief explanation of the energy system modelling analysis is provided in Annex 3. A full explanation is presented in the documentation for the Appraisal - volume 8 of Reference 10 - and the data used to describe renewable sources together with the detailed results of the analysis are presented in the Assessment of Renewable Energy for the UK (Ref 3).

Estimated contribution to energy supply *(table 3 in each module)*

For each technology the range of energy contribution is shown for the years 2005 and 2025. The range shown encompasses all six scenarios and all three discount regimes used in the Assessment (Ref 3).

Estimated carbon dioxide emissions savings *(table 3 in each module)*

For each technology the range of estimated carbon dioxide emission savings is shown for the years 2005 and 2025 based upon the estimated range of generation contribution.

The estimates of carbon dioxide savings assume that each unit of renewable energy generation effectively displaces a unit of conventional generation and its concomitant emissions. As it is not possible to say what specific conventional generation would be displaced by new renewable energy generation - it might be coal, gas, nuclear or other depending on the location and the level of uptake of renewables - these estimates assume that the carbon dioxide emissions displaced are equivalent to those from an average unit of plant on the 1990 UK electricity system. The emission savings can then be established by subtracting the net emissions produced by the renewable energy plant during operation from the displaced emissions.

Most renewables plant does not produce carbon dioxide during operation. The exceptions are those using biofuels. Of these, the biomass technologies - energy crops and agricultural and forestry wastes - produce no net carbon dioxide emissions during operation as the carbon is effectively being recycled. Wastes going to landfill generate methane gas which if not used for heat or power generation would need to be flared. It is therefore assumed that the use of landfill gas for energy production also results in no net carbon dioxide emissions. The situation is more complicated for the municipal and industrial wastes. A detailed consideration of the nature of these wastes and alternative disposal routes was required in order to derive the net emissions during operation and allow the calculation of emission savings relative to the average plant mix which are presented in this Annex.

Categorisation *(table 3 in most modules)*

Each new and renewable energy technology is assigned to one of three categories.

1.   Market enablement via NFFO and RDD&D.

2.   Assessment and RDD&D.

3.   Watching Brief

The categorisation of the technologies is based upon a consideration of all the information prepared during the course of the strategic review of new and renewable energy technologies.

Technology descriptions

The technology descriptions contained in this annex draw heavily upon the material published as Appendix 7 in the REAG report (Ref 2).

Glossary

Declared Net Capacity (DNC). The Declared Net Capacity for intermittent renewable plant is very broadly defined as the equivalent capacity of base load plant that would produce the same annual energy output.

EUREKA, LINK, SMART and SPUR are programmes promoted by the DTI.

ENERGY SYSTEM MODELLING - "MARKAL"

1. The contribution that any technology may make in the future will be determined principally by the availability of a commercial technology and of an exploitable resource, the economic competitiveness of the technology with other forms of generation (including those already supplying the system) and the demand for energy. In order to consider the role that particular technologies might play in a future UK energy market it is necessary to consider not only the new and renewables market but also the evolution of the whole UK energy market as developments here will determine the willingness of suppliers to enter and develop the new and renewables sector. This is notoriously difficult to predict and a range of views of the future, or scenarios, were therefore considered and their effect on the development of these technologies assessed.

2. The methodology adopted for the assessment utilised a scenario approach combined with energy system analysis using the IEA developed MARKAL (MARKet ALlocation) model. An important aspect of the study was to gauge how robust the estimated potential contributions of the technologies were to uncertainties regarding future international prices for primary energy and UK demands. This was achieved by running the model with prices and demands developed in scenarios which conceptualised very different economic and social backgrounds for the evolution of the UK and the international energy sectors. These scenarios, all of which were given equal weighting in the assessment, had the following themes.

i. **High Oil Price (HOP)**. This scenario envisaged a future in which oil and gas prices rise steeply to the highest levels considered sustainable, but coal prices remain low.

ii. **Composite (CSS)**. A scenario capturing a range of conventional thinking on the future development of prices and demands at the outset of the study in 1990.

iii. **Low Oil Price (LOP)**. A future in which oil and gas prices remain at their present low levels.

iv. **Heightened Environmental Concern A (HEC-a)**. A future in which society and economic management are strongly influenced by environmental concerns. This is manifest by a high carbon tax aimed at reducing carbon dioxide emissions coupled with other measures to encourage initiatives such as recycling and greater use of public transport. Also the scenario envisaged that their would be a moratorium on the construction of nuclear power plant and an early rundown of existing stations.

v. **Heightened Environmental Concern B (HEC-b)**. Essentially the same scenario as HEC-a except that additional nuclear development is permitted.

vi. **Shifting Sands (SS)**. Strictly, not a scenario in its own right but a sensitivity test imposed on the CSS scenario to investigate the effect which oil price "shocks" would have on the potential contribution of technologies.

3. It should be noted that these scenarios are not predictions of future UK energy prices and demands. The scenarios are "what if" tools, which provide a reference framework for assessing the prospects of the energy technologies and the robustness of these prospects to future uncertainties.

4. The analysis of the size and timing of the potential contributions for each scenario was made using the MARKAL model. The model is based on a linear programming system for minimising an "objective function", which in this case was the discounted cost of the overall energy system. The model yields a least cost solution for the energy system which gives the mix of technologies adopted to meet externally defined demands. For this study, the period 1995 to 2025 was considered, providing an assessment of the energy mix at 5 year intervals with results in terms of deployment against time for each technology considered - both conventional and new and renewable. Given the necessary database, the model can also quantify the environmental emissions associated with the energy system and can have emissions constraints imposed upon it in order to study the most cost effective technology responses to measures aimed at reducing environmental burdens. A full explanation of the energy systems model and the analysis methodology is presented in the documentation for the Appraisal of UK Energy RDD&D - Volume 8, Ref 10, and the data used to describe the renewable source together with the detailed results of the analysis are presented in the Assessment of Renewable Energy for the UK, Ref 3.

5. Generally the model operates with a single uniform discount rate applying to all technologies. For this assessment two such rates (8% and 15%) were investigated. These values were chosen as being reasonably representative of the rates required by major companies when considering main stream investments. In addition a "survey" case was investigated in which a 10% rate was assigned to main stream investments whilst a 25% rate was applied to non-core energy investments (e.g. energy efficiency) for a range of reasons such as availability of capital, information barriers, hidden costs etc.

6. The assessment of six scenarios against three different discounting regimes resulted in the consideration of 18 possible futures. In addition 8 further environmental sensitivity runs, based on the CSS and HOP scenarios, were undertaken. The total uptake of all electricity producing renewable energy technologies under these futures are shown in Figure 10.

Figure 10

Renewable Energy Electricity Generation under various MARKAL scenarios (including environmental scenarios)

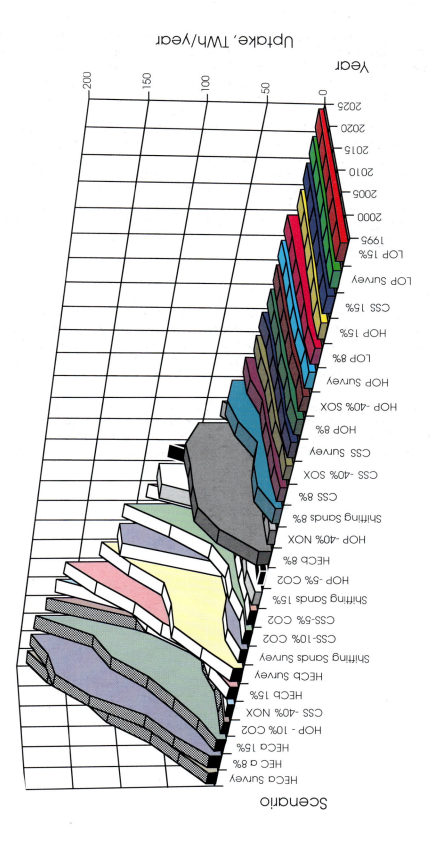

C4

Printed in the United Kingdom for HMSO.
Dd.297335, 3/94, C20, 3400, 5673, 282837.